Introduction

Chapter 1: The Role of Cities in Climate Change

Chapter 2: Smart Cities and Climate Resilience

Chapter 3: Renewable Energy Innovations

Chapter 4: Green Infrastructure and Urban Sustainability

Chapter 5: Transportation and Mobility Solutions

Chapter 6: Building Smart and Sustainable Cities

Chapter 7: Data-Driven Urban Planning

Chapter 8: Climate Adaptation Strategies for Cities

Chapter 9: Financing and Policy Frameworks

Conclusion

Introduction

Climate change and urbanization are two of the most significant challenges facing our world today. As the global population continues to grow, more people are moving into cities, leading to increased urbanization. At the same time, the impacts of climate change are becoming more pronounced, with rising temperatures, extreme weather events, and sea level rise posing significant threats to urban areas. Understanding the interplay between these two phenomena is crucial for developing effective strategies to mitigate and adapt to the impacts of climate change in urban environments.

Overview of Climate Change and Urbanization

The phenomenon of climate change is characterized by long-term alterations in temperature, precipitation patterns, and other climatic conditions on Earth. These changes are largely driven by human activities, particularly the emission of greenhouse gases (GHGs) such as carbon dioxide (CO_2), methane (CH_4), and nitrous oxide (N_2O). The burning of fossil fuels for energy, deforestation, and industrial processes are among the primary sources of these emissions. As a result, the concentration of GHGs in the atmosphere has increased dramatically since the industrial revolution, leading to a rise in global temperatures, commonly referred to as global warming.

Urbanization, on the other hand, refers to the increasing concentration of people living in cities. This trend has been accelerating over the past few decades, with more than half of the world's population now residing in urban areas. By 2050, it is projected that nearly 70% of the global population will live in cities. Urbanization is driven by various factors, including the search for better economic opportunities, improved living standards, and access to services such as healthcare and education. However, this rapid urban growth poses significant challenges for urban planning and sustainability.

The relationship between climate change and urbanization is complex and multifaceted. Cities are both major contributors to climate change and vulnerable to its impacts. Urban areas are responsible for a significant portion of global GHG emissions, primarily due to energy consumption for buildings, transportation, and industry. The high concentration of people and activities in cities leads to increased demand for energy, water, and other resources, resulting in higher emissions. Additionally, the urban heat island effect, where cities experience higher temperatures than their rural surroundings due to human activities and infrastructure, further exacerbates the impact of climate change.

At the same time, cities are highly vulnerable to the effects of climate change. Rising temperatures can lead to heatwaves, which pose health risks, particularly for vulnerable populations such as the elderly and those with preexisting health conditions. Extreme weather events, such as heavy rainfall and storms, can cause flooding, damaging infrastructure, disrupting transportation, and displacing residents. Sea level rise threatens coastal cities, leading to erosion, saltwater intrusion, and increased flooding risks. The combination of these factors can strain urban infrastructure and resources, making it challenging to maintain the quality of life for city residents.

Moreover, the impacts of climate change are not distributed evenly across cities. Socioeconomic factors play a significant role in determining a city's vulnerability to climate change. Low-income communities often lack the resources to adequately prepare for and respond to climate-related events, making them more susceptible to the adverse effects of climate change. This disparity highlights the need for equitable and inclusive approaches to urban planning and climate adaptation.

Addressing the challenges posed by climate change and urbanization requires a comprehensive and integrated approach. Cities must adopt strategies that promote sustainable development, reduce GHG emissions, and enhance resilience to climate impacts. This includes investing in renewable energy, improving energy efficiency, and implementing green infrastructure solutions. Additionally, urban planning must consider the long-term impacts of climate change, incorporating adaptive measures to protect vulnerable communities and ensure the sustainability of urban areas.

In conclusion, climate change and urbanization are interlinked challenges that require coordinated efforts to address. As urban populations continue to grow, cities must take proactive steps to mitigate their contributions to climate change and adapt to its impacts. By doing so, they can create sustainable, resilient urban environments that support the well-being of their residents and contribute to global efforts to combat climate change.

Importance of Technology and Innovation in Mitigating Climate Change

As the world grapples with the escalating threats of climate change, the role of technology and innovation becomes increasingly pivotal in mitigating its impacts. The rapid advancements in technology and the continuous flow of innovative ideas present unique opportunities to address the multifaceted challenges posed by climate change. This section explores the importance of technology and innovation in mitigating climate change and highlights the ways in which they contribute to a more sustainable and resilient future.

One of the primary ways technology aids in mitigating climate change is through the reduction of GHG emissions. Renewable energy technologies, such as solar, wind, and hydropower, play a crucial role in this effort. These technologies harness natural resources to generate electricity, reducing reliance on fossil fuels and cutting down carbon dioxide (CO_2) emissions. Solar panels, for instance, convert sunlight into electricity with minimal environmental impact, while wind turbines generate power from wind energy, providing a clean and sustainable energy source. The widespread adoption of these renewable energy technologies is essential for achieving global emissions reduction targets and transitioning to a low-carbon economy.

Energy efficiency is another critical area where technology and innovation have a significant impact. Advances in energy-efficient technologies, such as LED lighting, smart thermostats, and high-efficiency appliances, help reduce energy consumption in residential, commercial, and industrial sectors. Smart grids, which use digital technology to monitor and manage electricity usage, enable more efficient distribution of energy and reduce wastage. Innovations in building materials and design, such as insulation and green roofs, further contribute to energy savings by minimizing the need for heating and cooling. By improving energy efficiency, these technologies not only lower GHG emissions but also reduce energy costs for consumers and businesses.

Transportation is a major contributor to GHG emissions, and technological advancements in this sector are crucial for climate mitigation. Electric vehicles (EVs) are at the forefront of this transformation. Powered by electricity rather than fossil fuels, EVs produce zero tailpipe emissions and can be charged using renewable energy sources, significantly reducing their carbon footprint. Additionally, innovations in public transportation, such as electric buses and high-speed trains, offer sustainable alternatives to traditional, carbon-intensive modes of transport. The development of advanced battery technologies and the expansion of EV charging infrastructure are key to accelerating the adoption of electric mobility and achieving substantial emissions reductions in the transportation sector.

In addition to reducing emissions, technology and innovation are vital for climate adaptation and resilience. Urban areas, in particular, face heightened risks from climate change, including extreme weather events, sea-level rise, and heatwaves. Smart city technologies, which integrate information and communication technology (ICT) with urban infrastructure, enhance the resilience of cities by improving resource management and emergency response capabilities. For example, smart sensors can monitor air quality and weather conditions in real-time, enabling authorities to issue timely warnings and take preventive measures during extreme weather events. Similarly, advanced modeling and simulation tools help urban planners design resilient infrastructure that can withstand climate-related impacts.

Innovations in agriculture also play a crucial role in mitigating climate change. Precision agriculture technologies, such as drones, sensors, and data analytics, optimize the use of water, fertilizers, and pesticides, leading to more sustainable farming practices. These technologies not only enhance crop yields and food security but also reduce the environmental footprint of agriculture. Furthermore, advancements in biotechnology, such as the development of drought-resistant and high-yield crop varieties, contribute to climate resilience by enabling farmers to adapt to changing climatic conditions.

Carbon capture and storage (CCS) is another innovative approach to mitigating climate change. CCS technologies capture CO_2 emissions from industrial processes and power plants and store them underground, preventing them from entering the atmosphere. Although still in the early stages of deployment, CCS has the potential to significantly reduce emissions from hard-to-abate sectors, such as cement and steel production. Continued investment in research and development is essential to enhance the efficiency and scalability of CCS technologies.

Finally, technology and innovation drive public awareness and engagement in climate action. Digital platforms and social media enable widespread dissemination of information on climate change and sustainability, empowering individuals and communities to take action. Online tools and mobile apps provide resources for tracking and reducing personal carbon footprints, promoting sustainable lifestyle choices.

In conclusion, technology and innovation are indispensable in the fight against climate change. From renewable energy and energy efficiency to sustainable transportation and smart city solutions, technological advancements offer practical and scalable solutions to reduce GHG emissions and enhance climate resilience. Continued investment in research, development, and deployment of innovative technologies is crucial for achieving a sustainable and climate-resilient future.

Structure and Purpose of the Book

In the face of unprecedented climate change and rapid urbanization, cities around the world are at a crossroads. The decisions made today will determine the sustainability and livability of urban areas for future generations. This book, "Smart Cities and Climate Innovation: Harnessing Technology for Sustainable Urban Futures," aims to provide a comprehensive exploration of the pivotal role that technology and innovation play in addressing these dual challenges. The structure of this book is designed to guide readers through the complex interplay between urbanization and climate change, highlighting cutting-edge solutions and practical strategies for creating resilient, sustainable cities.

Purpose of the Book

The primary purpose of this book is to educate and inform policymakers, urban planners, engineers, architects, environmentalists, and other stakeholders about the critical importance of integrating technology and innovation into urban climate strategies. By providing a thorough understanding of the current landscape, emerging technologies, and innovative practices, this book seeks to inspire action and foster collaboration across sectors.

1. Raising Awareness: One of the key objectives is to raise awareness about the severity of climate change impacts on urban areas and the urgency of adopting innovative solutions. By presenting data, trends, and potential applications, the book aims to illustrate the pressing need for cities to adapt and mitigate climate risks.

2. Highlighting Technological Advancements: The book delves into various technological advancements that are instrumental in reducing greenhouse gas emissions, enhancing energy efficiency, and building climate resilience. It covers a broad spectrum of technologies, from renewable energy systems and smart grids to EVs and green infrastructure, providing readers with a comprehensive understanding of available tools and their applications.

3. Promoting Best Practices: Through detailed discussions of best practices and successful implementations, the book offers valuable insights into how cities can leverage technology and innovation to combat climate change. These insights serve as a source of inspiration and practical guidance for readers looking to implement similar initiatives in their own contexts.

4. Encouraging Collaborative Efforts: Addressing climate change in urban areas requires a coordinated effort among various stakeholders, including government agencies, private sector companies, non-profit organizations, and the general public. This book emphasizes the importance of collaboration and provides frameworks for effective partnerships and policy-making.

Structure of the Book

The book is organized into nine chapters, each focusing on a specific aspect of technology and innovation in the context of cities and climate change. Here is a detailed breakdown of the structure:

Chapter 1: The Role of Cities in Climate Change: This chapter sets the stage by examining the unique position of cities as both contributors to and victims of climate change. It explores urbanization trends, energy consumption patterns, and the specific vulnerabilities of urban areas to climate impacts.

Chapter 2: Smart Cities and Climate Resilience: Focusing on the concept of smart cities, this chapter discusses how integrating ICT with urban infrastructure can enhance resilience. Topics include smart grids, water management, waste management, and the circular economy.

Chapter 3: Renewable Energy Innovations: This chapter delves into the latest advancements in renewable energy technologies, including solar, wind, and hydropower. It examines how these technologies can be integrated into urban planning to reduce reliance on fossil fuels and lower carbon emissions.

Chapter 4: Green Infrastructure and Urban Sustainability: Green infrastructure plays a crucial role in urban sustainability. This chapter explores various green solutions, such as green roofs, urban forests, and sustainable urban drainage systems, highlighting their benefits and implementation strategies.

Chapter 5: Transportation and Mobility Solutions: Transportation is a significant source of urban emissions. This chapter covers innovations in electric vehicles, public transportation, bike-sharing programs, and strategies to reduce traffic congestion and promote sustainable mobility.

Chapter 6: Building Smart and Sustainable Cities: The focus of this chapter is on creating energy-efficient buildings and retrofitting existing structures to meet modern sustainability standards. It discusses smart building technologies, sustainable construction practices, and relevant policies and incentives.

Chapter 7: Data-Driven Urban Planning: Data and predictive analytics are powerful tools for urban planners. This chapter explores how big data and the Internet of Things (IoT) can be used to inform urban planning decisions, improve resource management, and engage citizens in participatory planning processes.

Chapter 8: Climate Adaptation Strategies for Cities: Adaptation is crucial for cities facing climate risks. This chapter provides strategies for mitigating urban heat islands, managing floods and coastal protection, and designing resilient infrastructure to withstand climate impacts.

Chapter 9: Financing and Policy Frameworks: Implementing technological and innovative solutions requires substantial financial investments and supportive policy frameworks. This chapter discusses funding mechanisms, public-private partnerships, and international and national policy initiatives that facilitate climate action in urban areas.

Conclusion: The conclusion summarizes the key points discussed in the book, emphasizing the importance of continued investment in technology and innovation. It calls to action policymakers, practitioners, and citizens to collaborate in building a sustainable and resilient future for cities.

By following this structure, the book aims to provide a holistic view of how technology and innovation can be harnessed to mitigate climate change and promote sustainable urban development. Each chapter builds on the previous one, creating a coherent narrative that guides readers through the complexities of urban climate action and equips them with the knowledge and tools needed to make a positive impact.

Chapter 1: The Role of Cities in Climate Change

Urban areas are at the forefront of both the causes and the impacts of climate change. As hubs of economic activity, innovation, and population growth, cities are responsible for a significant portion of global GHG emissions. This chapter explores the intricate relationship between urbanization and climate change, highlighting how cities contribute to and are affected by these environmental changes.

We begin by examining the trends in urbanization and how the rapid growth of cities worldwide has led to increased energy consumption and higher emissions. Understanding these patterns is crucial for developing strategies to curb emissions and promote sustainable urban development. Next, we delve into the specific vulnerabilities of urban areas to climate change, including extreme weather events, sea level rise, and the urban heat island effect. These challenges underscore the urgency of integrating climate resilience into urban planning and policy-making.

By the end of this chapter, readers will have a comprehensive understanding of the pivotal role cities play in the climate change narrative. This foundation sets the stage for subsequent chapters, which will explore innovative solutions and technological advancements aimed at mitigating the impacts of climate change on urban environments.

Urbanization Trends and Climate Impact

The 21st century has witnessed an unprecedented surge in urbanization, with more than half of the world's population now residing in urban areas. This trend is projected to continue, with estimates suggesting that nearly 70% of the global population will live in cities by 2050. This rapid urban growth is driven by a multitude of factors, including economic opportunities, improved living standards, and better access to healthcare, education, and other essential services. However, urbanization also poses significant challenges, particularly in the context of climate change. This section delves into the trends of urbanization and their consequent impact on climate, emphasizing the need for sustainable urban planning and development.

Urbanization Trends

The global urban population has been steadily increasing over the past few decades. In 1950, only about 30% of the world's population lived in urban areas. By 2018, this figure had risen to 55%, and it is expected to reach 68% by 2050. This growth is not uniform across all regions; developing countries, particularly in Asia and Africa, are experiencing the most rapid urbanization. Cities like Lagos, Kinshasa, Mumbai, and Jakarta are among the fastest-growing urban areas in the world. This explosive growth is largely driven by rural-to-urban migration, as individuals and families move to cities in search of better economic prospects and living conditions.

Urbanization brings several benefits, including economic growth, innovation, and cultural exchange. Cities are often seen as engines of economic development, providing jobs, education, and healthcare. The concentration of resources and people in urban areas can lead to increased productivity and innovation, driving advancements in technology and industry. Furthermore, cities offer diverse cultural and social opportunities, contributing to the overall quality of life.

However, the rapid and often unplanned expansion of urban areas can lead to numerous challenges, including inadequate infrastructure, housing shortages, and increased pollution. As cities grow, they consume more land, water, and energy, putting significant pressure on natural resources and ecosystems. The environmental footprint of urban areas is substantial, contributing to climate change and exacerbating its impacts.

Climate Impact of Urbanization

Urban areas are significant contributors to GHG emissions, which are the primary drivers of climate change. Cities consume about 75% of the world's energy and are responsible for approximately 70% of global carbon dioxide (CO_2) emissions. The high concentration of buildings, vehicles, and industrial activities in urban areas leads to increased energy consumption and, consequently, higher emissions.

Energy Consumption and Emissions

The energy demand in cities is primarily driven by residential, commercial, and industrial activities. Buildings, for instance, require heating, cooling, lighting, and other electrical services, which consume significant amounts of energy. Transportation is another major source of emissions, with the proliferation of vehicles contributing to air pollution and GHG emissions. Industrial activities, including manufacturing and construction, also add to the energy demand and emissions in urban areas.

The reliance on fossil fuels for energy generation in many cities exacerbates the emission of GHGs. Coal, oil, and natural gas are the primary sources of energy in urban areas, contributing to the high levels of CO_2 and other pollutants in the atmosphere. While some cities are transitioning to renewable energy sources, the pace of this transition is slow compared to the rate of urbanization and the growing energy demand.

Urban Heat Island Effect

The urban heat island (UHI) effect is a phenomenon where urban areas experience higher temperatures than their rural surroundings. This effect is primarily caused by human activities and the modification of land surfaces. Buildings, roads, and other infrastructure absorb and retain heat, leading to higher temperatures in cities. The lack of vegetation and green spaces in urban areas further exacerbates the UHI effect.

The UHI effect not only contributes to higher energy consumption for cooling but also poses health risks, particularly during heatwaves. Increased temperatures can lead to heat-related illnesses and exacerbate existing health conditions, particularly among vulnerable populations such as the elderly and those with preexisting health conditions.

Air Pollution

Urbanization leads to increased air pollution, with harmful pollutants such as particulate matter (PM), nitrogen oxides (NOx), and sulfur dioxide (SO2) being released into the atmosphere. These pollutants are primarily generated from transportation, industrial activities, and energy production. Poor air quality in cities can have severe health impacts, including respiratory and cardiovascular diseases.

Additionally, air pollution contributes to climate change by enhancing the greenhouse effect. For instance, black carbon, a component of particulate matter, absorbs sunlight and heats the atmosphere, contributing to global warming.

Land Use and Ecosystem Degradation

The expansion of urban areas often leads to the conversion of natural landscapes into built environments. This process results in the loss of forests, wetlands, and other critical ecosystems that play a vital role in regulating the climate. Deforestation for urban development reduces the planet's capacity to absorb CO2, further contributing to climate change. The degradation of ecosystems also affects biodiversity, with many species losing their natural habitats. Urban sprawl and the encroachment into natural areas disrupt ecosystems and diminish their resilience to climate impacts.

The Need for Sustainable Urban Development

Given the significant impact of urbanization on climate change, it is crucial to adopt sustainable urban development practices. Cities must prioritize reducing their carbon footprint by transitioning to renewable energy sources, improving energy efficiency, and promoting sustainable transportation options. Urban planning should incorporate green infrastructure, such as parks, green roofs, and urban forests, to mitigate the UHI effect and enhance resilience to climate impacts.

Furthermore, policies and regulations should be implemented to control air pollution and promote sustainable land use practices. This includes preserving natural landscapes, promoting compact and efficient urban growth, and enhancing public transportation systems. Collaboration between government agencies, private sector companies, and communities is essential to achieve these goals and create sustainable, livable urban environments.

Energy Consumption and Emissions in Urban Areas

Urban areas are the epicenters of modern economic activities, social interactions, and technological advancements. However, they are also significant contributors to energy consumption and GHG emissions, making them central to the discussions on climate change mitigation. This section delves into the dynamics of energy consumption in urban areas, the sources of emissions, and the implications for climate change.

Energy Consumption in Urban Areas

Urban areas account for a substantial portion of global energy consumption. The high density of buildings, infrastructure, industries, and transportation systems in cities drives significant energy demand. This energy consumption can be broadly categorized into three main sectors: residential, commercial, and industrial.

Residential Sector

In urban areas, residential energy consumption is driven by various needs, including heating, cooling, lighting, and powering appliances. The demand for energy in homes is influenced by factors such as building design, insulation quality, the efficiency of heating and cooling systems, and the types of appliances used.

Heating and cooling are particularly energy-intensive activities. In colder climates, significant amounts of energy are used to heat homes during winter, while in hotter climates, air conditioning is a major energy consumer. Poorly insulated buildings and inefficient heating and cooling systems exacerbate energy consumption.

Commercial Sector

The commercial sector, which includes offices, retail spaces, restaurants, and other service-oriented establishments, also contributes significantly to urban energy demand. Similar to the residential sector, energy is used for heating, cooling, lighting, and operating various electronic devices and equipment.

Large commercial buildings, such as skyscrapers and shopping malls, require substantial energy for HVAC (heating, ventilation, and air conditioning) systems, elevators, escalators, and extensive lighting. The rise of digital technologies and the proliferation of electronic devices have further increased energy consumption in the commercial sector.

Industrial Sector

Industrial activities in urban areas, including manufacturing, processing, and construction, are major energy consumers. These activities often involve energy-intensive processes that require large amounts of electricity and heat.

Industries such as steel production, chemical manufacturing, and cement production are particularly energy-intensive. The machinery, equipment, and processes used in these industries consume significant amounts of energy, contributing to the overall energy demand in urban areas.

Sources of Emissions in Urban Areas

The high energy consumption in urban areas is primarily met through the burning of fossil fuels, leading to significant GHG emissions. These emissions are a major driver of climate change. The primary sources of emissions in urban areas include the following.

Electricity Generation

A substantial portion of urban energy consumption is supplied by electricity, which is often generated from fossil fuels such as coal, natural gas, and oil. The combustion of these fuels releases large amounts of CO_2 and other GHGs into the atmosphere.

Coal-fired power plants are among the highest emitters of CO_2 due to the carbon-intensive nature of coal. Natural gas, while cleaner than coal, still produces significant emissions. The shift towards renewable energy sources, such as wind, solar, and hydropower, is crucial for reducing emissions from electricity generation.

Transportation

Transportation is a major source of urban emissions, driven by the widespread use of vehicles powered by gasoline and diesel. The internal combustion engines in these vehicles emit CO_2, methane (CH_4), nitrous oxide (N_2O), and other pollutants.

Urban areas often face significant traffic congestion, leading to idling engines and increased fuel consumption. Public transportation systems, while more efficient than private vehicles, also contribute to emissions, particularly if they rely on fossil fuels. The transition to EVs and the expansion of public transportation networks powered by renewable energy can significantly reduce transportation-related emissions.

Industrial Processes

Industrial activities in urban areas contribute to emissions through the combustion of fossil fuels for energy and the release of process-related emissions. Industries such as cement, steel, and chemical manufacturing are particularly notable for their high emissions.

Cement production, for instance, involves the calcination of limestone, which releases CO_2. Similarly, the chemical reactions involved in steel production emit GHGs. Improving energy efficiency and adopting cleaner technologies in industrial processes are critical for reducing emissions.

Building Operations

The operation of residential, commercial, and institutional buildings contributes to urban emissions. Buildings consume energy for heating, cooling, lighting, and powering various devices and systems. This energy is often sourced from fossil fuel-based power plants, leading to indirect emissions. The materials used in building construction, such as concrete and steel, also have a significant carbon footprint. Implementing energy-efficient building designs, retrofitting existing structures, and using sustainable construction materials can help mitigate emissions from buildings.

Implications for Climate Change

The high energy consumption and associated emissions in urban areas have profound implications for climate change. The concentration of GHGs in the atmosphere leads to global warming, which in turn drives a range of climate-related impacts, including extreme weather events, sea level rise, and changes in precipitation patterns. Urban areas are particularly vulnerable to these impacts due to their high population densities and critical infrastructure.

Heatwaves

The urban heat island effect exacerbates the impact of heatwaves in cities. The increased temperatures in urban areas can lead to higher mortality rates, especially among vulnerable populations such as the elderly and those with preexisting health conditions. Increased energy demand for air conditioning during heatwaves further strains energy systems and increases emissions.

Flooding

Urban areas are prone to flooding due to their extensive impervious surfaces, such as roads and buildings, which prevent natural absorption of rainwater. Climate change-induced extreme weather events, including heavy rainfall and storms, can overwhelm urban drainage systems, leading to severe flooding. Flooding can damage infrastructure, disrupt transportation, and displace residents.

Air Quality

The emissions from transportation, industrial activities, and power generation contribute to poor air quality in urban areas. Pollutants such as particulate matter (PM), nitrogen oxides (NOx), and sulfur dioxide (SO2) have detrimental health effects, including respiratory and cardiovascular diseases. Climate change can exacerbate air quality issues by increasing the frequency and intensity of heatwaves, which enhance the formation of ground-level ozone, a harmful air pollutant.

The Path Forward

Addressing the challenges of energy consumption and emissions in urban areas requires a multifaceted approach that includes policy interventions, technological advancements, and behavioral changes. Cities must prioritize the transition to renewable energy sources, enhance energy efficiency across all sectors, and promote sustainable transportation options. Urban planning should incorporate green infrastructure to mitigate the urban heat island effect and improve resilience to climate impacts.

Renewable Energy

Expanding the use of renewable energy sources, such as solar, wind, and hydropower, is crucial for reducing emissions from electricity generation. Incentives and policies that support the deployment of renewable energy technologies can accelerate this transition.

Energy Efficiency

Improving energy efficiency in buildings, transportation, and industrial processes can significantly reduce energy consumption and emissions. This includes adopting energy-efficient appliances, retrofitting buildings with better insulation, and optimizing industrial processes for lower energy use.

Sustainable Transportation

Promoting public transportation, cycling, walking, and the adoption of EVs can reduce emissions from the transportation sector. Investments in public transportation infrastructure and EV charging networks are essential for supporting sustainable mobility.

Green Infrastructure

Integrating green spaces, such as parks, green roofs, and urban forests, into city planning can help mitigate the urban heat island effect, improve air quality, and enhance the resilience of urban areas to climate impacts.

Vulnerabilities of Cities to Climate Change

Cities, as hubs of human activity and economic development, face significant vulnerabilities to the impacts of climate change. These vulnerabilities stem from their high population densities, extensive infrastructure, and the concentration of economic activities. Climate change exacerbates existing urban challenges and introduces new risks, posing threats to the health, safety, and livelihoods of urban residents. This section explores the various vulnerabilities of cities to climate change, including extreme weather events, sea level rise, urban heat islands, and social inequalities.

Extreme Weather Events

Climate change is increasing the frequency and intensity of extreme weather events, such as hurricanes, storms, heavy rainfall, and heatwaves. Cities, with their dense populations and extensive infrastructure, are particularly vulnerable to these events.

Flooding and Storm Surges

Heavy rainfall and storms can overwhelm urban drainage systems, leading to flooding. The impermeable surfaces in cities, such as roads and buildings, prevent water from being absorbed into the ground, exacerbating flood risks. Flooding can cause significant damage to homes, businesses, and infrastructure, disrupt transportation networks, and displace residents.

Coastal cities are especially at risk from storm surges, which are intensified by rising sea levels. Storm surges can inundate coastal areas, causing severe flooding, erosion, and damage to coastal infrastructure.

Heatwaves

Heatwaves are becoming more frequent and severe due to climate change. Urban areas, with their high concentrations of buildings, roads, and other heat-absorbing surfaces, are particularly susceptible to the urban heat island effect. This phenomenon leads to higher temperatures in cities compared to surrounding rural areas.

Heatwaves pose serious health risks, particularly to vulnerable populations such as the elderly, young children, and those with preexisting health conditions. Increased temperatures can lead to heat exhaustion, heatstroke, and exacerbate respiratory and cardiovascular diseases. The demand for air conditioning during heatwaves also strains energy systems, potentially leading to power outages.

Sea Level Rise

Rising sea levels, driven by the melting of polar ice caps and thermal expansion of seawater, pose a significant threat to coastal cities. Sea level rise exacerbates the risk of flooding, erosion, and saltwater intrusion.

Flooding and Erosion

Coastal cities are at risk of both gradual inundation and sudden flooding events due to sea level rise. Low-lying areas are particularly vulnerable, with the potential for permanent submersion of land. This can lead to the displacement of communities, loss of property, and damage to infrastructure. Erosion of coastlines can undermine buildings, roads, and other infrastructure located near the shore. The loss of beaches and coastal habitats also affects tourism and local economies.

Saltwater Intrusion

Rising sea levels can cause saltwater to intrude into freshwater aquifers, contaminating drinking water supplies. This is a critical issue for coastal cities that rely on groundwater for their water supply. Saltwater intrusion can also damage agricultural lands, reducing their productivity and impacting food security.

Urban Heat Island Effect

The UHI effect is a phenomenon where urban areas experience higher temperatures than their rural surroundings. This effect is primarily caused by human activities and the modification of land surfaces.

Increased Energy Demand

Higher temperatures in cities lead to increased demand for air conditioning and cooling, straining energy systems and increasing greenhouse gas emissions. The higher energy consumption during peak periods can also lead to power outages, affecting homes, businesses, and essential services.

Health Impacts

The UHI effect exacerbates the health impacts of heatwaves, leading to higher mortality and morbidity rates. Vulnerable populations, including the elderly, children, and those with chronic illnesses, are particularly at risk. Prolonged exposure to high temperatures can lead to heat-related illnesses, dehydration, and exacerbation of existing health conditions.

Environmental Degradation

The UHI effect contributes to environmental degradation by increasing the demand for water and energy, leading to higher emissions and resource depletion. The lack of green spaces and vegetation in urban areas also reduces biodiversity and the ability of cities to mitigate the impacts of climate change.

Social Inequalities

Climate change exacerbates existing social inequalities in urban areas. Vulnerable populations, including low-income communities, marginalized groups, and informal settlers, are disproportionately affected by climate impacts.

Housing and Infrastructure

Low-income communities often reside in substandard housing that is more susceptible to damage from extreme weather events and flooding. These areas may lack adequate infrastructure, such as drainage systems and emergency services, increasing their vulnerability to climate impacts.

Access to Resources

Vulnerable populations may have limited access to resources such as clean water, healthcare, and emergency services. During climate-related disasters, these communities may struggle to recover and rebuild, leading to prolonged displacement and hardship.

Economic Impacts

Climate change can exacerbate economic inequalities by disproportionately affecting sectors such as agriculture, fishing, and tourism, which many low-income communities rely on for their livelihoods. Loss of income and employment opportunities due to climate impacts can increase poverty and reduce resilience.

The Need for Resilience and Adaptation

Given the vulnerabilities of cities to climate change, it is crucial to develop and implement strategies to enhance urban resilience and adaptation. This involves a multifaceted approach that includes improving infrastructure, enhancing emergency preparedness, promoting social equity, and integrating climate considerations into urban planning.

Infrastructure Upgrades

Investing in resilient infrastructure, such as flood defenses, green roofs, and permeable pavements, can help cities cope with extreme weather events and reduce the impacts of the UHI effect. Upgrading drainage systems and implementing nature-based solutions can mitigate flood risks and improve water management.

Emergency Preparedness

Enhancing emergency preparedness and response capabilities is essential for protecting urban populations during climate-related disasters. This includes developing early warning systems, evacuation plans, and emergency shelters. Community engagement and education are also critical for building resilience at the local level.

Social Equity

Addressing social inequalities is key to ensuring that all urban residents can cope with and recover from climate impacts. Policies and programs that improve access to housing, healthcare, and essential services for vulnerable populations are vital. Empowering communities through participatory planning and decision-making processes can also enhance resilience.

Climate-Informed Urban Planning

Integrating climate considerations into urban planning and development is crucial for creating resilient cities. This includes adopting land use policies that reduce exposure to climate risks, promoting sustainable transportation, and incorporating green infrastructure into urban designs.

Chapter 2: Smart Cities and Climate Resilience

As urbanization continues to accelerate, cities face increasing challenges related to climate change, including extreme weather events, resource scarcity, and environmental degradation. In response, the concept of smart cities has emerged as a transformative approach to urban development, leveraging advanced technologies and data-driven solutions to enhance sustainability and resilience. This chapter delves into the critical role of smart cities in building climate resilience, exploring how integrating ICT with urban infrastructure can mitigate climate risks and improve the quality of life for urban residents.

We begin by defining what constitutes a smart city and the key components that differentiate it from traditional urban environments. The chapter then examines the various technologies that underpin smart city initiatives, including smart grids, water management systems, and waste management solutions. Each of these technologies plays a vital role in reducing greenhouse gas emissions, optimizing resource use, and enhancing urban resilience to climate impacts.

Additionally, this chapter highlights the importance of the circular economy in the context of smart cities. By promoting the reuse, recycling, and regeneration of resources, cities can reduce their environmental footprint and create more sustainable urban ecosystems. The integration of circular economy principles with smart city technologies presents a powerful strategy for addressing climate change and fostering long-term sustainability.

Through this exploration, readers will gain a comprehensive understanding of how smart cities can serve as a blueprint for climate resilience. The insights provided will not only illustrate the potential of technological innovation in mitigating climate risks but also underscore the importance of strategic planning, policy support, and community engagement in realizing the vision of smart, sustainable cities.

Definition and Components of Smart Cities

A smart city is an urban area that utilizes ICT and other advanced technologies to enhance the quality of life for its residents, improve the efficiency of urban services, and promote sustainability. By integrating various technological solutions, smart cities aim to address the challenges posed by rapid urbanization and climate change, creating resilient and sustainable urban environments.

Definition of Smart Cities

A smart city leverages ICT to collect, analyze, and act upon data in real time, enabling better decision-making and more efficient management of resources. The goal is to create a city that is more responsive to the needs of its citizens, improves operational efficiency, reduces environmental impact, and fosters economic development. Key characteristics of a smart city include the following.

Connectivity and Data Integration

Smart cities are characterized by their ability to connect various devices, systems, and services through the IoT. This interconnected network allows for seamless data collection and sharing across different urban domains, such as transportation, energy, water, and waste management.

Sustainability

Sustainability is a core principle of smart cities. By using technology to optimize resource use, reduce waste, and lower greenhouse gas emissions, smart cities aim to create environmentally friendly urban environments that can adapt to and mitigate the impacts of climate change.

Citizen-Centric Services

Smart cities prioritize the needs and well-being of their residents. By providing efficient and accessible public services, enhancing public safety, and fostering community engagement, smart cities strive to improve the overall quality of life for their inhabitants.

Components of Smart Cities

The components of smart cities can be categorized into several key areas, each playing a vital role in creating a sustainable and resilient urban ecosystem.

Smart Infrastructure

Smart infrastructure incorporates advanced digital technologies to enhance the efficiency, reliability, and sustainability of urban systems:

- Smart Grids: Advanced electrical grids that use digital technology to monitor and manage the distribution of electricity. Smart grids improve energy efficiency, reduce outages, and facilitate the integration of renewable energy sources.
- Smart Buildings: Buildings equipped with sensors and automation systems to optimize energy use, enhance security, and improve comfort. Smart buildings can adjust lighting, heating, and cooling based on occupancy and environmental conditions.
- Smart Transportation: Intelligent transportation systems that use real-time data to manage traffic flow, reduce congestion, and promote the use of public transit. This includes smart traffic lights, autonomous vehicles, and integrated mobility platforms.

Smart Environment

Smart environment technologies leverage advanced systems to monitor and manage natural resources, ensuring sustainability and enhancing urban livability:

- Water Management: Technologies that monitor and manage water supply, distribution, and quality. Smart water meters, leak detection systems, and automated irrigation help conserve water and ensure its efficient use.
- Waste Management: Systems that optimize waste collection, recycling, and disposal. Smart waste bins with sensors can alert waste management services when they need to be emptied, reducing unnecessary collections and improving efficiency.
- Air Quality Monitoring: Networks of sensors that track air pollution levels in real-time. Data from these sensors can inform policies to reduce emissions and protect public health.

Smart Governance

Smart governance utilizes digital tools and data analytics to enhance transparency, efficiency, and citizen engagement in government operations:

- E-Government Services: Digital platforms that facilitate interaction between citizens and government. These services include online portals for paying bills, accessing public records, and reporting issues, making government more accessible and transparent.
- Data-Driven Decision Making: The use of data analytics and artificial intelligence (AI) to inform urban planning and policy decisions. By analyzing trends and patterns, cities can make more informed choices that improve efficiency and sustainability.

Smart Economy

A smart economy leverages technology and innovation to drive sustainable economic growth and inclusivity:

- Innovation Hubs: Centers that foster innovation, entrepreneurship, and collaboration among businesses, academia, and government. These hubs support the development of new technologies and solutions that drive economic growth and sustainability.
- Digital Inclusion: Efforts to ensure that all citizens have access to digital technologies and the skills needed to participate in the digital economy. This includes providing affordable internet access and digital literacy programs.

Smart Living

Smart living integrates advanced technologies to improve the quality of life, safety, and well-being of urban residents:

- Public Safety: Technologies that enhance the safety and security of urban environments. This includes surveillance systems, emergency response coordination, and predictive policing tools.
- Healthcare: Smart healthcare systems that use telemedicine, remote monitoring, and electronic health records to improve patient care and access to medical services.
- Education: Digital learning platforms and smart classrooms that provide personalized education and improve learning outcomes.

Smart Grids and Energy Efficiency

In the quest for sustainable urban development, smart grids and energy efficiency play a crucial role in transforming how cities consume and manage energy. As urban populations grow and energy demands increase, traditional power grids face significant challenges in ensuring reliable and efficient energy distribution. Smart grids, leveraging advanced technologies and real-time data, offer a solution by enhancing the resilience, efficiency, and sustainability of urban energy systems. This section explores the concept of smart grids, their components, and their role in promoting energy efficiency in cities.

Understanding Smart Grids

A smart grid is an electricity network that uses digital communication technology to detect and react to local changes in usage. This advanced grid system integrates various technologies, including smart meters, sensors, and data analytics, to enhance the reliability, efficiency, and sustainability of electricity distribution. Unlike traditional grids, which operate on a one-way communication model, smart grids facilitate two-way communication between the utility and its customers, enabling real-time monitoring and management of energy flows.

Key components of smart grids include:

- Smart Meters: Smart meters are digital devices that record electricity consumption in real time and communicate this information to both the utility and the consumer. These meters provide detailed insights into energy usage patterns, enabling consumers to make informed decisions about their energy consumption and allowing utilities to optimize grid operations.
- Advanced Sensors: Sensors deployed throughout the grid infrastructure monitor various parameters, such as voltage levels, current flows, and equipment status. These sensors help detect faults, predict equipment failures, and ensure the efficient operation of the grid.
- Automated Control Systems: Automated control systems use real-time data to manage grid operations dynamically. These systems can reroute power flows, balance supply and demand, and respond to outages or other anomalies quickly, improving grid reliability and resilience.

- Energy Storage Solutions: Energy storage systems, such as batteries, are integrated into smart grids to store excess energy generated during periods of low demand and release it during peak demand. This capability helps stabilize the grid and ensures a steady supply of electricity.
- Renewable Energy Integration: Smart grids facilitate the integration of renewable energy sources, such as solar and wind power, by managing the variability and intermittency of these energy sources. Advanced forecasting and grid management tools enable the efficient use of renewable energy, reducing reliance on fossil fuels.

Enhancing Energy Efficiency through Smart Grids

Smart grids contribute to energy efficiency in several ways, from reducing energy losses and optimizing energy consumption to enabling demand response programs and supporting renewable energy integration. These improvements not only lower greenhouse gas emissions but also reduce energy costs for consumers and utilities.

Reducing Energy Losses

Traditional power grids suffer from significant energy losses during transmission and distribution due to outdated infrastructure and inefficiencies. Smart grids, with their advanced monitoring and control capabilities, can identify and mitigate these losses. By optimizing power flows and reducing the need for long-distance transmission, smart grids enhance overall energy efficiency.

Optimizing Energy Consumption

Smart meters provide consumers with detailed information about their energy usage, enabling them to identify and eliminate wasteful practices. Utilities can also use this data to develop targeted energy efficiency programs, such as offering incentives for energy-efficient appliances or promoting off-peak energy use. These measures help balance supply and demand, reducing strain on the grid and lowering energy consumption.

Demand Response Programs

Demand response programs incentivize consumers to reduce or shift their energy usage during peak demand periods. Smart grids enable real-time communication between utilities and consumers, allowing for dynamic pricing and automated load control. For example, smart thermostats can adjust heating and cooling settings based on grid conditions, reducing peak demand and enhancing grid stability.

Supporting Renewable Energy Integration

The variability and intermittency of renewable energy sources pose challenges for traditional grids. Smart grids, with their advanced forecasting and real-time management capabilities, can efficiently integrate renewable energy into the grid. By matching energy supply with demand and storing excess renewable energy, smart grids ensure a stable and reliable energy supply while maximizing the use of clean energy sources.

Enhanced Grid Resilience

Smart grids improve the resilience of the electricity network by enabling faster detection and response to outages and other disruptions. Automated systems can isolate fault areas and reroute power, minimizing the impact of outages on consumers. Enhanced resilience reduces the need for backup power generation, which is often less efficient and more polluting than primary sources.

Case Studies and Examples

Several cities around the world have implemented smart grid initiatives, demonstrating the potential benefits of this technology. For example:

Amsterdam, Netherlands

Amsterdam has developed a comprehensive smart grid project that includes the installation of smart meters, the integration of renewable energy sources, and the deployment of EVs charging stations. The project aims to reduce energy consumption, lower carbon emissions, and enhance the city's overall energy resilience.

Austin, Texas, USA

The Pecan Street Project in Austin is a leading example of a smart grid initiative. The project involves advanced energy storage solutions, demand response programs, and real-time energy monitoring. Residents participate in energy efficiency programs and use smart home technologies to optimize their energy consumption, resulting in significant energy savings and reduced emissions.

Songdo, South Korea

Songdo is a purpose-built smart city that incorporates smart grid technologies from the ground up. The city features smart meters, automated control systems, and renewable energy integration. These technologies ensure efficient energy use, reduce greenhouse gas emissions, and enhance the quality of life for residents.

Challenges and Future Directions

While the benefits of smart grids are clear, there are several challenges to their widespread adoption. These include the high initial costs of infrastructure upgrades, the need for regulatory and policy support, and concerns about data privacy and cybersecurity. Addressing these challenges requires coordinated efforts from governments, utilities, and technology providers:

- Cost and Infrastructure: Upgrading traditional power grids to smart grids involves significant investment in infrastructure and technology. Financial incentives and funding support

from governments can help utilities overcome these barriers and accelerate the transition to smart grids.
- Regulatory and Policy Support: Effective regulatory frameworks and policies are essential to promote the adoption of smart grids. This includes setting standards for smart grid technologies, providing incentives for energy efficiency, and supporting renewable energy integration.
- Data Privacy and Cybersecurity: The digital nature of smart grids raises concerns about data privacy and cybersecurity. Ensuring robust security measures and protecting consumer data are critical for building trust and ensuring the safe operation of smart grids.

Water Management and Smart Technology

Water management is a critical component of urban sustainability, particularly in the face of climate change and rapid urbanization. Cities around the world are grappling with challenges related to water scarcity, flooding, and water quality. Smart technology offers innovative solutions to enhance water management, making urban areas more resilient, efficient, and sustainable. This section explores the application of smart technology in water management, focusing on its benefits, components, and examples of successful implementation.

Importance of Water Management

Effective water management is essential for ensuring the availability of clean water, protecting public health, and maintaining the ecological balance. Urban areas face several water-related challenges, including:

- Water Scarcity: Rapid urbanization and population growth increase the demand for water, often outstripping the available supply. Climate change exacerbates this issue by altering precipitation patterns and reducing the reliability of water sources.

- Flooding: Urban areas are prone to flooding due to the high concentration of impervious surfaces, such as roads and buildings, which prevent natural absorption of rainwater. Extreme weather events, intensified by climate change, further increase the risk of urban flooding.
- Water Quality: Industrial activities, agricultural runoff, and inadequate wastewater treatment can contaminate water sources, posing significant health risks to urban populations. Ensuring water quality is crucial for public health and environmental protection.

Smart Technology in Water Management

Smart technology offers a range of tools and systems to address these challenges, improve water management practices, and enhance the resilience of urban areas. Key components of smart water management include the following

Smart Water Meters

Smart water meters provide real-time data on water consumption, enabling utilities and consumers to monitor usage patterns, detect leaks, and identify inefficiencies. These meters transmit data wirelessly, allowing for remote monitoring and management.

Advanced Sensors

Sensors deployed throughout the water infrastructure monitor various parameters, such as flow rates, pressure, and water quality. These sensors help detect anomalies, predict equipment failures, and ensure the efficient operation of the water supply system.

Automated Control Systems

Automated control systems use real-time data to manage water distribution dynamically. These systems can adjust water flow, pressure, and treatment processes based on current conditions, improving efficiency and reducing waste.

Data Analytics

Data analytics tools process the vast amounts of data collected by smart meters and sensors, providing actionable insights for water management. These tools help utilities optimize operations, predict demand, and plan for future water needs.

Internet of Things (IoT)

The IoT connects various devices and systems within the water infrastructure, enabling seamless communication and coordination. This interconnected network facilitates real-time monitoring and management of water resources, enhancing overall system efficiency.

Benefits of Smart Water Management

The integration of smart technology in water management offers numerous benefits, including the following.

Improved Efficiency

Smart water management systems optimize the use of water resources by identifying and eliminating inefficiencies. Real-time monitoring and automated controls reduce water losses, minimize energy consumption, and ensure the effective distribution of water.

Enhanced Resilience

By providing real-time data and predictive analytics, smart technology enhances the resilience of urban water systems to climate-related impacts. Early detection of leaks and anomalies allows for timely interventions, reducing the risk of system failures and water shortages.

Better Water Quality

Advanced sensors and monitoring tools ensure the continuous assessment of water quality, enabling prompt detection and remediation of contamination. This helps protect public health and maintain the ecological integrity of water bodies.

Cost Savings

Efficient water management reduces operational costs for utilities and lowers water bills for consumers. The early detection of leaks and inefficiencies prevents costly repairs and conserves valuable water resources.

Informed Decision-Making

Data analytics provides utilities with detailed insights into water usage patterns, demand trends, and system performance. This information supports informed decision-making and strategic planning for future water needs.

Examples of Smart Water Management

Several cities around the world have successfully implemented smart water management systems, demonstrating the potential benefits of this technology. Notable examples include:

Singapore

Singapore's national water agency, PUB, has integrated smart technology into its water management strategy. The city-state uses smart water meters, advanced sensors, and data analytics to monitor water consumption, detect leaks, and ensure water quality. PUB's smart water grid enables efficient water distribution and enhances resilience to water scarcity.

Barcelona, Spain

Barcelona has implemented a comprehensive smart water management system that includes smart meters, sensors, and IoT technology. The city's smart water network monitors water consumption in real time, detects leaks, and optimizes water distribution. This system has significantly reduced water losses and improved the efficiency of the city's water supply.

Amsterdam, Netherlands

Amsterdam's smart water management initiatives focus on flood prevention and water quality. The city uses advanced sensors and automated control systems to monitor water levels, flow rates, and water quality in its canals and rivers. These technologies help manage flood risks, ensure water quality, and enhance the city's resilience to climate change.

San Francisco, USA

San Francisco Public Utilities Commission (SFPUC) has adopted smart water meters and data analytics to monitor and manage water consumption. The city's advanced metering infrastructure provides detailed insights into water usage, enabling consumers to track their consumption and detect leaks. SFPUC's smart water management system has led to significant water savings and improved operational efficiency.

Challenges and Future Directions

While smart technology offers significant benefits for water management, there are several challenges to its widespread adoption. These include the high initial costs of implementation, the need for regulatory support, and concerns about data privacy and cybersecurity. Addressing these challenges requires coordinated efforts from governments, utilities, and technology providers.

Cost and Infrastructure

Upgrading traditional water infrastructure to incorporate smart technology involves substantial investment. Financial incentives and funding support from governments can help utilities overcome these barriers and accelerate the transition to smart water management.

Regulatory and Policy Support

Effective regulatory frameworks and policies are essential to promote the adoption of smart water management systems. This includes setting standards for smart water technologies, providing incentives for efficient water use, and supporting research and development.

Data Privacy and Cybersecurity

The digital nature of smart water management raises concerns about data privacy and cybersecurity. Ensuring robust security measures and protecting consumer data are critical for building trust and ensuring the safe operation of smart water systems.

Waste Management and Circular Economy

Effective waste management is a critical aspect of urban sustainability, directly impacting environmental health, resource efficiency, and the quality of life for urban residents. Traditional linear waste management systems, which follow a "take, make, dispose" model, are becoming increasingly unsustainable due to the growing volume of waste generated by urban populations. In contrast, the circular economy offers a transformative approach to waste management by promoting the reuse, recycling, and regeneration of materials. This section explores the integration of smart technology in waste management and the principles of the circular economy, highlighting their benefits, components, and examples of successful implementation.

The Challenge of Waste Management in Urban Areas

Urban areas generate a significant amount of waste due to high population densities, economic activities, and consumption patterns. Effective waste management is essential for preventing environmental pollution, conserving resources, and maintaining public health. However, cities face several challenges in managing waste, including:

- Increasing Waste Volumes: Rapid urbanization and population growth lead to higher waste generation. Managing this growing volume of waste requires efficient collection, transportation, treatment, and disposal systems.
- Environmental Impact: Improper waste management can result in environmental pollution, including soil, water, and air contamination. Landfills and incineration, common waste disposal methods, contribute to greenhouse gas emissions and other environmental issues.
- Resource Depletion: The traditional linear waste management model leads to the depletion of natural resources, as materials are discarded rather than reused or recycled. This unsustainable approach contributes to resource scarcity and environmental degradation.

Smart Technology in Waste Management

Smart technology offers innovative solutions to enhance the efficiency and sustainability of waste management systems. By leveraging data, automation, and connectivity, smart waste management systems optimize waste collection, improve recycling rates, and reduce environmental impact. Key components of smart waste management include the following

Smart Waste Bins

Smart waste bins are equipped with sensors that monitor fill levels and send real-time data to waste management operators. These bins can alert operators when they need to be emptied, optimizing collection routes and reducing unnecessary collections. This technology improves operational efficiency, reduces fuel consumption, and minimizes traffic congestion.

Automated Sorting Systems

Automated sorting systems use advanced technologies such as robotics, artificial intelligence (AI), and machine learning to sort recyclable materials from mixed waste streams. These systems improve the accuracy and efficiency of recycling processes, increasing the recovery rates of valuable materials and reducing contamination.

Data Analytics

Data analytics tools process information collected from smart waste bins, sorting systems, and other sources to provide insights into waste generation patterns, recycling rates, and system performance. This data-driven approach enables waste management operators to optimize operations, plan for future needs, and identify opportunities for improvement.

Internet of Things (IoT)

The IoT connects various devices and systems within the waste management infrastructure, enabling seamless communication and coordination. This interconnected network facilitates real-time monitoring and management of waste collection, transportation, and processing, enhancing overall system efficiency.

The Circular Economy: Principles and Benefits

The circular economy is a regenerative system that aims to minimize waste and make the most of resources. Unlike the traditional linear economy, the circular economy keeps products, materials, and resources in use for as long as possible through principles such as designing for longevity, reusing, recycling, and regenerating natural systems. Key principles of the circular economy include:

- Designing for Longevity: Products are designed to last longer, with a focus on durability, reparability, and upgradability. This approach reduces the need for frequent replacements and minimizes waste generation.
- Promoting Reuse: The circular economy encourages the reuse of products and materials through practices such as refurbishment, remanufacturing, and repurposing. By extending the lifecycle of products, cities can reduce waste and conserve resources.
- Enhancing Recycling: Recycling plays a crucial role in the circular economy by transforming waste materials into new products. Efficient recycling systems ensure that valuable materials are recovered and reintroduced into the production cycle, reducing the need for virgin resources.
- Regenerating Natural Systems: The circular economy aims to restore and regenerate natural systems by returning valuable nutrients to the soil and minimizing environmental impact. This approach supports biodiversity, enhances ecosystem services, and promotes sustainable agriculture.

Examples of Circular Economy in Waste Management

Several cities around the world have embraced the principles of the circular economy in their waste management strategies, demonstrating the potential benefits of this approach. Notable examples include the following.

Copenhagen, Denmark

Copenhagen has implemented a comprehensive waste management system that incorporates smart technology and circular economy principles. The city's waste-to-energy plants convert non-recyclable waste into electricity and district heating, reducing landfill use and greenhouse gas emissions. Additionally, Copenhagen promotes recycling and composting through extensive public education and infrastructure.

San Francisco, USA

San Francisco aims to achieve zero waste by 2025 through its innovative waste management programs. The city has implemented mandatory recycling and composting, incentivized the use of reusable products, and invested in advanced sorting facilities. These efforts have significantly reduced the city's landfill waste and increased recycling rates.

Tokyo, Japan

Tokyo's 23 wards have adopted an advanced waste management system that includes smart waste bins, automated sorting facilities, and extensive recycling programs. The city has also promoted the circular economy by encouraging businesses to design products for durability and reparability, reducing waste generation at the source.

Challenges and Future Directions

While the integration of smart technology and circular economy principles in waste management offers significant benefits, there are several challenges to their widespread adoption. These include the high initial costs of implementation, the need for regulatory support, and the importance of public engagement and behavior change.

Cost and Infrastructure

Upgrading traditional waste management systems to incorporate smart technology and circular economy practices requires substantial investment. Financial incentives and funding support from governments can help municipalities overcome these barriers and accelerate the transition to sustainable waste management.

Regulatory and Policy Support

Effective regulatory frameworks and policies are essential to promote the adoption of smart waste management systems and circular economy practices. This includes setting standards for waste management technologies, providing incentives for recycling and reuse, and supporting research and development.

Public Engagement

Public engagement and behavior change are crucial for the success of smart waste management and circular economy initiatives. Educating residents about the benefits of recycling, reuse, and waste reduction, and encouraging sustainable consumption patterns, can drive participation and support for these initiatives.

Chapter 3: Renewable Energy Innovations

As the world grapples with the urgent need to address climate change, renewable energy innovations have emerged as a cornerstone of sustainable urban development. Traditional energy systems, heavily reliant on fossil fuels, are major contributors to greenhouse gas emissions and environmental degradation. Transitioning to renewable energy sources is essential for reducing emissions, enhancing energy security, and creating resilient urban environments. This chapter delves into the latest advancements in renewable energy technologies, exploring how cities can harness these innovations to build a sustainable future.

We begin by examining the various types of renewable energy sources, including solar, wind, and hydropower, and their potential applications in urban settings. The chapter then explores cutting-edge technologies that enhance the efficiency and integration of these energy sources into the urban grid. From advanced solar panels and wind turbines to innovative energy storage solutions and smart grid integration, these technologies are transforming the urban energy landscape.

Additionally, this chapter highlights the importance of policy support, financial incentives, and community engagement in promoting the adoption of renewable energy. Successful case studies from cities around the world will illustrate the practical implementation of these technologies, demonstrating their potential to drive urban sustainability and climate resilience.

By the end of this chapter, readers will have a comprehensive understanding of the role of renewable energy innovations in shaping the future of urban energy systems. The insights provided will underscore the critical importance of embracing renewable energy to achieve long-term sustainability and mitigate the impacts of climate change.

Solar Energy Technologies

Solar energy technologies have emerged as one of the most promising solutions for meeting the growing energy demands of urban areas while mitigating the impacts of climate change. Harnessing the power of the sun, these technologies offer a clean, renewable, and abundant source of energy. This section explores the various solar energy technologies, their applications in urban settings, and the advancements that are driving their adoption and efficiency.

Photovoltaic (PV) Systems

Photovoltaic (PV) systems, commonly known as solar panels, are the most widely recognized solar energy technology. These systems convert sunlight directly into electricity using semiconductor materials. The basic unit of a PV system is the solar cell, which is made of silicon or other semiconducting materials. When sunlight strikes the solar cell, it excites electrons, creating an electric current.

Types of PV Systems

Types of PV systems include:

- Monocrystalline Silicon Panels: These panels are made from single-crystal silicon and are known for their high efficiency and durability. They have a higher power output per square meter compared to other types.
- Polycrystalline Silicon Panels: Made from multiple silicon crystals, these panels are less efficient than monocrystalline but are cheaper to produce.
- Thin-Film Solar Cells: These cells are made by depositing one or more thin layers of photovoltaic material onto a substrate. They are lightweight and flexible, making them suitable for a variety of applications, including building-integrated photovoltaics (BIPV).

Applications in Urban Settings

PV systems can be installed on rooftops, facades, and other structures, making them highly versatile for urban environments. Rooftop solar installations are particularly popular for residential and commercial buildings, allowing property owners to generate their own electricity and reduce their reliance on the grid. Building-integrated photovoltaics (BIPV) incorporate solar panels into the building materials, such as windows, walls, and roofs. This integration not only generates electricity but also enhances the aesthetic appeal of buildings.

Advancements in PV Technology

Recent advancements in PV technology have focused on improving efficiency and reducing costs. Innovations such as bifacial solar panels, which capture sunlight on both sides of the panel, and perovskite solar cells, which offer high efficiency at a lower cost, are driving the next generation of PV systems.

Solar Thermal Systems

Solar thermal systems use sunlight to produce heat, which can be used for various applications such as water heating, space heating, and power generation. These systems are classified into two main categories: passive and active solar thermal systems.

Passive Solar Heating

Passive solar heating involves designing buildings to collect, store, and distribute solar heat without the use of mechanical systems. This is achieved through architectural elements such as south-facing windows, thermal mass materials (e.g., concrete, brick), and natural ventilation. Passive solar design reduces the need for artificial heating and cooling, thereby lowering energy consumption and costs.

Active Solar Heating

Active solar heating systems use mechanical equipment, such as pumps and fans, to circulate heat-absorbing fluids (e.g., water, air) through solar collectors. These collectors capture and transfer solar heat to storage systems or directly to the building's heating system. Common applications include solar water heaters, which use solar collectors to heat water for domestic use, and solar space heating systems, which provide heating for buildings.

Concentrated Solar Power (CSP)

CSP systems use mirrors or lenses to concentrate sunlight onto a small area, generating high temperatures that can be used to produce steam and drive turbines for electricity generation. These systems are typically used in large-scale power plants but can also be adapted for urban applications. CSP technologies include parabolic troughs, solar power towers, and linear Fresnel reflectors, each with its unique method of concentrating sunlight.

Solar Energy Storage

One of the key challenges of solar energy is its intermittent nature, as it is only available during daylight hours. Energy storage solutions are essential for ensuring a reliable and continuous power supply.

Battery Storage

Battery storage systems store excess electricity generated by solar PV systems during the day for use during nighttime or cloudy periods. Lithium-ion batteries are the most common type used for residential and commercial solar installations due to their high energy density and efficiency. Advancements in battery technology, such as improved battery chemistry and energy management systems, are enhancing the capacity, lifespan, and affordability of solar energy storage.

Thermal Storage

Thermal storage systems store solar heat for later use. For instance, solar thermal power plants can use molten salt to store heat, which can be converted into electricity when needed. Residential systems can store hot water in insulated tanks for domestic use.

Grid Integration and Virtual Power Plants

Integrating solar energy with the grid allows for excess power to be fed into the grid and distributed to other users. Virtual power plants (VPPs) aggregate distributed energy resources, including solar PV systems and storage, to create a flexible and reliable power supply.

Benefits of Solar Energy Technologies

The adoption of solar energy technologies offers numerous benefits for urban areas:

- Environmental Benefits: Solar energy is a clean and renewable resource, reducing greenhouse gas emissions and dependence on fossil fuels. It helps mitigate climate change and improve air quality in cities.
- Economic Benefits: Solar installations can reduce electricity bills for homeowners and businesses, provide energy independence, and create jobs in the renewable energy sector. Incentives and subsidies can further enhance the economic attractiveness of solar investments.
- Energy Security: By diversifying the energy mix and reducing reliance on imported fuels, solar energy enhances energy security and resilience against energy price fluctuations and supply disruptions.
- Scalability and Flexibility: Solar energy technologies can be scaled to meet different energy needs, from small residential systems to large-scale solar farms. Their flexibility makes them suitable for a wide range of applications in urban environments.

Wind Energy Developments

Wind energy has become one of the fastest-growing sources of renewable energy worldwide, driven by advancements in technology and growing concerns about climate change and energy security. As cities seek to reduce their carbon footprint and diversify their energy sources, wind energy offers a viable solution with numerous environmental and economic benefits. This section explores the latest developments in wind energy technology, its applications in urban settings, and the challenges and opportunities associated with its adoption.

The Basics of Wind Energy

Wind energy harnesses the kinetic energy of moving air to generate electricity. Wind turbines convert this kinetic energy into mechanical power, which is then converted into electrical power through a generator. The two primary types of wind turbines are horizontal-axis wind turbines (HAWTs) and vertical-axis wind turbines (VAWTs).

Horizontal-Axis Wind Turbines (HAWTs)

HAWTs are the most common type of wind turbines, featuring a rotor with blades that rotate around a horizontal axis. These turbines are typically installed in large wind farms, both onshore and offshore, and are known for their high efficiency and power output.

Vertical-Axis Wind Turbines (VAWTs)

VAWTs have a rotor that rotates around a vertical axis. These turbines are less common but offer advantages in certain urban applications due to their ability to capture wind from any direction and their lower noise levels.

Technological Advancements in Wind Energy

Significant advancements in wind energy technology have improved the efficiency, reliability, and cost-effectiveness of wind turbines, making them more attractive for urban and rural installations.

Increased Turbine Size and Efficiency

Modern wind turbines have grown significantly in size, with larger rotor diameters and taller towers. This increase in size allows turbines to capture more wind energy, resulting in higher power output. Innovations in blade design, such as longer and lighter blades made from advanced composite materials, have also enhanced turbine efficiency.

Offshore Wind Energy

Offshore wind farms take advantage of stronger and more consistent winds available at sea. Advances in offshore wind technology, including floating wind turbines and improved foundations, have expanded the potential for offshore wind energy. Floating turbines can be installed in deeper waters, increasing the available area for wind energy development.

Energy Storage Integration

Integrating energy storage solutions with wind turbines helps address the intermittency of wind energy. Battery storage systems can store excess electricity generated during periods of high wind and release it during low-wind periods, ensuring a stable and reliable power supply. Advancements in battery technology, such as longer lifespan and higher energy density, are making this integration more feasible.

Smart Grid and Digitalization

The incorporation of smart grid technologies and digitalization has enhanced the management and optimization of wind energy. Advanced sensors, data analytics, and machine learning algorithms enable real-time monitoring and predictive maintenance of wind turbines, reducing downtime and operational costs. These technologies also facilitate better integration of wind energy into the grid.

Urban Applications of Wind Energy

While large-scale wind farms are typically located in rural or offshore areas, there is growing interest in urban wind energy applications. Urban wind turbines can provide a supplementary energy source for cities, contributing to their sustainability goals.

Building-Integrated Wind Turbines

Building-integrated wind turbines (BIWTs) are small-scale turbines installed on or within buildings. These turbines can be incorporated into the architectural design of skyscrapers, office buildings, and residential complexes. BIWTs can generate electricity for the building, reducing its reliance on the grid and lowering energy costs.

Rooftop Wind Turbines

Rooftop wind turbines are another option for urban wind energy generation. These small-scale turbines can be installed on the roofs of homes, schools, and commercial buildings. They are particularly suitable for areas with consistent wind patterns and can complement rooftop solar panels.

Public Infrastructure and Urban Spaces

Wind turbines can also be installed in public spaces such as parks, plazas, and along highways. These installations can serve dual purposes, generating clean energy and raising public awareness about renewable energy. Urban wind projects can be designed to blend aesthetically with the surroundings, minimizing visual impact.

Challenges and Opportunities

The adoption of wind energy in urban areas presents several challenges and opportunities that need to be addressed to maximize its potential.

Noise and Aesthetic Concerns

One of the primary challenges of urban wind energy is the noise generated by wind turbines. Advances in turbine design, such as quieter blades and improved gearboxes, have mitigated this issue to some extent. Additionally, integrating turbines into building designs and urban landscapes can address aesthetic concerns.

Variable Wind Patterns

Urban areas often experience variable and turbulent wind patterns due to the presence of buildings and other structures. Designing turbines that can efficiently capture wind energy in such conditions is crucial. VAWTs are particularly suited for urban environments as they can harness wind from any direction and perform well in turbulent conditions.

Regulatory and Zoning Barriers

Urban wind energy projects may face regulatory and zoning barriers, such as height restrictions and permitting requirements. Policymakers and urban planners need to develop clear guidelines and streamlined processes to facilitate the deployment of urban wind turbines.

Community Engagement and Support

Gaining community support is essential for the success of urban wind energy projects. Engaging with residents, addressing their concerns, and highlighting the environmental and economic benefits of wind energy can build public acceptance and support.

Case Studies and Examples

Several cities have successfully integrated wind energy into their urban environments, demonstrating the potential of this renewable energy source.

London, UK

The Strata SE1 building in London features three integrated wind turbines on its roof. These turbines provide a portion of the building's electricity needs and serve as a visible commitment to sustainability.

New York City, USA

The Brooklyn Navy Yard is home to a small wind turbine that generates electricity for the industrial park. This project showcases the feasibility of urban wind energy in densely populated areas.

Rotterdam, Netherlands

The city of Rotterdam has installed several small wind turbines on public buildings and infrastructure. These installations contribute to the city's renewable energy goals and enhance its reputation as a leader in sustainability.

Hydropower and Urban Applications

Hydropower is a well-established and highly efficient renewable energy source that harnesses the energy of flowing or falling water to generate electricity. Traditionally associated with large-scale projects like dams and reservoirs, hydropower can also be applied in urban environments through innovative technologies and design strategies. This section explores the basics of hydropower, the advancements that enable its use in urban areas, and the potential benefits and challenges of urban hydropower applications.

Basics of Hydropower

Hydropower works by converting the kinetic energy of water into mechanical energy using turbines, which then drive generators to produce electricity. The three main types of hydropower systems are:

Impoundment Hydropower

This involves the creation of a reservoir using a dam to store a large quantity of water. Electricity is generated by releasing water from the reservoir through turbines. While highly effective, this type of hydropower is generally unsuitable for urban environments due to the need for extensive space and significant environmental impact.

Run-of-River Hydropower

Run-of-river systems generate electricity by channeling a portion of a river's flow through a turbine without the need for large reservoirs. These systems have a smaller environmental footprint and are more adaptable to urban settings.

Pumped Storage Hydropower

This method involves pumping water to a higher elevation during periods of low electricity demand and releasing it to generate electricity during peak demand periods. It functions as a form of energy storage but is less commonly implemented in urban areas due to space requirements.

Urban Applications of Hydropower

Innovations in technology and urban design have enabled the integration of hydropower into city landscapes, providing a sustainable energy solution that can complement other renewable energy sources. Key urban hydropower applications include:

Micro-Hydropower Systems

Micro-hydropower systems are small-scale installations that generate up to 100 kW of electricity. These systems can be integrated into urban water infrastructure, such as stormwater management systems, irrigation channels, and drinking water supply networks. They provide a decentralized energy source suitable for urban environments.

In-Pipe Hydropower

In-pipe hydropower technology involves installing turbines within existing water pipelines to generate electricity from the flow of water. This approach is particularly effective in cities with extensive water distribution networks, as it can generate power without requiring additional infrastructure.

Low-Head Hydropower

Low-head hydropower systems are designed to operate on small drops in water elevation, making them suitable for urban rivers, canals, and weirs. These systems can generate electricity with minimal impact on water flow and local ecosystems.

Tidal and Wave Energy

Urban coastal areas can harness tidal and wave energy to generate electricity. Tidal energy systems capture the kinetic energy of tidal currents, while wave energy systems convert the energy of ocean waves. Both methods can be integrated into urban waterfronts and harbors, providing a reliable source of renewable energy.

Benefits of Urban Hydropower

Urban hydropower offers several advantages that make it an attractive option for sustainable city planning:

- Renewable and Clean Energy: Hydropower is a renewable energy source that produces no direct greenhouse gas

emissions during operation. It contributes to reducing the carbon footprint of urban areas and helps mitigate climate change.
- Energy Efficiency: Hydropower systems have high energy conversion efficiencies, often exceeding 90%. This efficiency makes them a reliable and effective source of electricity for urban applications.
- Decentralized Energy Production: Integrating hydropower into urban water infrastructure enables decentralized energy production, reducing the strain on centralized power grids and enhancing energy security.
- Multi-Functional Infrastructure: Urban hydropower systems can be integrated into existing water management infrastructure, such as pipelines and stormwater systems, providing dual benefits of energy generation and improved water management.

Challenges and Considerations

Despite the benefits, there are several challenges and considerations associated with urban hydropower:

- Environmental Impact: While generally lower than large-scale hydropower projects, urban hydropower systems can still impact local ecosystems and water quality. Careful design and environmental assessments are necessary to minimize these effects.
- Space and Infrastructure Requirements: Implementing hydropower in urban areas requires suitable water flow and infrastructure. Not all cities have the necessary conditions for effective hydropower generation.
- Regulatory and Permitting Issues: Urban hydropower projects must navigate complex regulatory environments and obtain necessary permits, which can be time-consuming and costly.
- Maintenance and Operational Challenges: Ensuring the reliability and efficiency of urban hydropower systems

requires regular maintenance and monitoring, which can pose logistical challenges in dense urban settings.

Integration of Renewable Energy in Urban Planning

Integrating renewable energy into urban planning is essential for creating sustainable, resilient, and low-carbon cities. As urban areas continue to grow, the demand for energy increases, making it crucial to adopt strategies that reduce greenhouse gas emissions and promote the use of renewable energy sources. This section explores the importance of incorporating renewable energy into urban planning, the strategies for achieving this integration, and the benefits and challenges associated with it.

Importance of Integrating Renewable Energy in Urban Planning

Urban areas are significant consumers of energy, accounting for a substantial portion of global energy demand and carbon emissions. Integrating renewable energy into urban planning is crucial for several reasons:

- Reducing Carbon Emissions: The transition to renewable energy sources, such as solar, wind, and hydropower, helps reduce carbon emissions from fossil fuel-based energy generation. This shift is vital for mitigating climate change and improving air quality in cities.
- Enhancing Energy Security: Renewable energy sources diversify the energy supply and reduce dependence on imported fuels. This enhances energy security and resilience against energy price volatility and supply disruptions.
- Promoting Sustainable Development: Integrating renewable energy into urban planning supports sustainable development by promoting the use of clean energy, reducing environmental impact, and fostering economic growth through the creation of green jobs.

Strategies for Integrating Renewable Energy

Successful integration of renewable energy into urban planning requires a multi-faceted approach that includes policy support, innovative design, and community engagement. Key strategies include:

Incorporating Renewable Energy in Building Codes and Standards

Implementing building codes and standards that mandate or incentivize the use of renewable energy systems, such as solar panels and wind turbines, in new constructions and renovations. This ensures that buildings are designed to maximize energy efficiency and utilize renewable energy sources.

Designing Energy-Efficient Urban Infrastructure

Planning urban infrastructure with energy efficiency in mind, including the use of smart grids, energy storage systems, and district heating and cooling networks. These systems can integrate renewable energy sources and optimize energy distribution and consumption.

Promoting Mixed-Use Development

Encouraging mixed-use development that combines residential, commercial, and industrial areas. This approach reduces the need for long commutes, lowers transportation energy consumption, and provides opportunities for localized renewable energy generation.

Utilizing Urban Spaces for Renewable Energy

Leveraging urban spaces, such as rooftops, facades, and public areas, for the installation of renewable energy systems. Rooftop solar panels, building-integrated photovoltaics (BIPV), and small-scale wind turbines can generate clean energy within the city.

Supporting Community Energy Projects

Encouraging community energy projects that involve local residents and businesses in the planning, financing, and operation of renewable energy systems. These projects foster community engagement, enhance local energy resilience, and create economic opportunities.

Implementing Green Infrastructure

Integrating green infrastructure, such as green roofs, urban forests, and permeable pavements, into urban planning. These features not only enhance energy efficiency but also provide co-benefits such as improved air quality, stormwater management, and increased biodiversity.

Benefits of Integrating Renewable Energy in Urban Planning

The integration of renewable energy into urban planning offers numerous benefits for cities and their residents:

- Environmental Benefits: Reducing reliance on fossil fuels lowers carbon emissions and air pollutants, contributing to a healthier urban environment and mitigating climate change.
- Economic Benefits: Investing in renewable energy creates jobs in the green economy, stimulates local economic growth, and can lead to cost savings on energy bills for households and businesses.
- Social Benefits: Access to clean energy enhances the quality of life for urban residents by providing reliable, affordable, and sustainable energy. Community energy projects also foster social cohesion and local empowerment.
- Resilience and Adaptability: Renewable energy systems enhance the resilience of urban areas to energy supply disruptions and climate impacts. Decentralized energy generation and storage systems provide a reliable energy supply during emergencies.

Challenges and Considerations

While the integration of renewable energy into urban planning offers significant benefits, several challenges must be addressed:

Initial Costs and Financing

The upfront costs of renewable energy systems and infrastructure can be high. Access to financing, incentives, and subsidies is crucial to overcome this barrier and promote widespread adoption.

Regulatory and Policy Barriers

Inconsistent regulations and policies can hinder the integration of renewable energy. Streamlined permitting processes, supportive policies, and clear guidelines are essential for facilitating renewable energy projects.

Technical and Logistical Challenges

Implementing renewable energy systems in densely populated urban areas can be technically and logistically challenging. Innovative design and engineering solutions are required to integrate these systems seamlessly into the urban fabric.

Public Acceptance and Engagement

Gaining public support and engagement is critical for the success of renewable energy projects. Educating residents about the benefits and addressing concerns through transparent communication and participatory planning processes can build public trust and acceptance.

Chapter 4: Green Infrastructure and Urban Sustainability

As cities around the world grapple with the challenges of rapid urbanization and climate change, the concept of green infrastructure has emerged as a vital strategy for promoting urban sustainability. Green infrastructure refers to a network of natural and semi-natural systems and spaces that provide environmental, economic, and social benefits. These include parks, green roofs, urban forests, rain gardens, and other elements that help manage stormwater, reduce urban heat, improve air quality, and enhance biodiversity.

This chapter explores the multifaceted role of green infrastructure in fostering sustainable urban development. It delves into the various types of green infrastructure and their applications in urban settings, highlighting their benefits for cities and their residents. We will examine how green infrastructure can mitigate the impacts of climate change, enhance urban resilience, and contribute to the overall well-being of urban populations.

By integrating green infrastructure into urban planning and design, cities can create healthier, more livable environments that support both people and nature. Through case studies and examples from cities that have successfully implemented green infrastructure initiatives, this chapter aims to provide readers with practical insights and inspiration for incorporating green solutions into their urban landscapes.

Green Roofs and Walls

Green roofs and walls are innovative components of green infrastructure that play a significant role in promoting urban sustainability. By integrating vegetation into buildings, these systems provide a range of environmental, economic, and social benefits, enhancing the quality of urban life and contributing to the resilience of cities. This section explores the different types of green roofs and walls, their benefits, implementation considerations, and examples of successful applications in urban environments.

Types of Green Roofs

Green roofs, also known as vegetated or living roofs, consist of a layer of vegetation planted over a waterproofing system installed on top of a flat or slightly sloped roof. There are two main types of green roofs: extensive and intensive.

Extensive Green Roofs

Extensive green roofs are characterized by their lightweight construction and shallow soil depth, typically ranging from 2 to 6 inches. These roofs are designed to be low-maintenance and support drought-tolerant plants such as sedums, grasses, and mosses. Extensive green roofs are ideal for large-scale applications where structural load capacity is limited, making them suitable for residential buildings, commercial properties, and industrial facilities.

Intensive Green Roofs

Intensive green roofs, also known as rooftop gardens, have deeper soil layers that can support a wider variety of plants, including shrubs, trees, and even small urban farms. Soil depths for intensive roofs can range from 6 inches to several feet. These roofs require more maintenance, irrigation, and structural support due to their increased weight. Intensive green roofs offer greater design flexibility and can create functional recreational spaces for building occupants.

Types of Green Walls

Green walls, or living walls, are vertical structures covered with vegetation. They can be attached to the exterior or interior walls of buildings. There are two primary types of green walls: green facades and living walls.

Green Facades

Green facades involve growing climbing plants directly on a building's surface or on a support structure such as a trellis or wire mesh. The plants root in the ground or in containers at the base of the wall and climb upward. This type of green wall is relatively simple to install and maintain, making it a cost-effective option for adding greenery to buildings.

Living Walls

Living walls, also known as vertical gardens, consist of pre-planted panels or modular systems attached to a wall. These panels contain a growing medium, such as soil or hydroponic substrates, and are equipped with irrigation and drainage systems. Living walls can support a diverse range of plant species, including herbs, flowers, and ferns. They require more complex installation and maintenance compared to green facades but offer greater aesthetic and functional benefits.

Benefits of Green Roofs and Walls

The integration of green roofs and walls into urban environments offers numerous benefits:

Environmental Benefits

Environmental benefits include:

- Stormwater Management: Green roofs and walls absorb and retain rainwater, reducing runoff and mitigating the risk of urban flooding. They also help filter pollutants from rainwater, improving water quality.
- Urban Heat Island Mitigation: Vegetation cools the surrounding air through evapotranspiration and shading, reducing the urban heat island effect and lowering temperatures in cities.

- Improved Air Quality: Plants capture airborne pollutants and particulate matter, enhancing air quality and reducing respiratory issues for urban residents.
- Biodiversity Enhancement: Green roofs and walls provide habitats for birds, insects, and other wildlife, supporting urban biodiversity.

Economic Benefits

Economic benefits include:

- Energy Efficiency: Green roofs and walls act as natural insulators, reducing the need for heating in winter and cooling in summer. This leads to energy savings and lower utility bills for building owners.
- Extended Roof Lifespan: The protective layer of vegetation shields the underlying roof membrane from UV radiation, temperature fluctuations, and mechanical damage, extending its lifespan and reducing maintenance costs.
- Increased Property Value: Buildings with green roofs and walls are often more attractive to tenants and buyers, potentially increasing property values and occupancy rates.

Social Benefits

Social benefits include:

- Enhanced Aesthetics: Green roofs and walls improve the visual appeal of urban environments, creating more pleasant and inviting spaces for residents and visitors.
- Health and Well-Being: Access to green spaces has been linked to improved mental health, reduced stress levels, and increased physical activity. Green roofs and walls provide opportunities for urban agriculture and community gardening, promoting healthy lifestyles.

- Noise Reduction: Vegetation on green roofs and walls can help absorb and deflect sound, reducing noise pollution in densely populated urban areas.

Implementation Considerations

While the benefits of green roofs and walls are clear, several factors must be considered to ensure their successful implementation

Structural Support

The building must be assessed for its load-bearing capacity to support the additional weight of a green roof or wall, including the soil, plants, and water retention systems. Structural reinforcements may be necessary for older buildings or those with limited load capacity.

Waterproofing and Drainage

Proper waterproofing is essential to prevent water damage to the building. Green roofs require a waterproof membrane, root barrier, and drainage layer to manage water flow and protect the structure. Green walls need an effective irrigation and drainage system to ensure plant health and prevent water seepage.

Plant Selection

Choosing the right plants is critical for the success of green roofs and walls. Plants must be selected based on their suitability to local climate conditions, light availability, and maintenance requirements. Drought-tolerant and native species are often preferred for their resilience and ecological benefits.

Maintenance

Regular maintenance is necessary to ensure the health and functionality of green roofs and walls. This includes irrigation, fertilization, pruning, and monitoring for pests and diseases. Maintenance requirements vary depending on the type of green roof or wall and the plant species used.

Examples of Successful Applications

Examples of successful applications include the following.

The High Line, New York City, USA:

The High Line is a renowned example of green infrastructure, transforming a disused elevated railway into a linear park with extensive green roofing and walls. This urban oasis provides recreational space, supports biodiversity, and mitigates the urban heat island effect.

Bosco Verticale, Milan, Italy

Bosco Verticale (Vertical Forest) is a pair of residential towers featuring extensive green walls with over 900 trees and 20,000 plants. This project demonstrates the potential for integrating substantial green infrastructure into high-rise buildings, enhancing urban biodiversity and air quality.

City Hall, Chicago, USA

The green roof on Chicago's City Hall covers over 20,000 square feet and includes a variety of native plants. This installation reduces stormwater runoff, lowers energy consumption, and serves as a model for sustainable urban design.

Urban Forests and Green Spaces

Urban forests and green spaces are critical components of sustainable urban environments. They provide a myriad of ecological, economic, and social benefits that enhance the quality of life for city residents and contribute to urban resilience. As cities face the challenges of climate change, air pollution, and increasing population densities, the integration of green spaces becomes even more essential. This section explores the importance of urban forests and green spaces, their benefits, strategies for implementation, and examples of successful urban green initiatives.

Importance of Urban Forests and Green Spaces

Urban forests consist of trees, shrubs, and other vegetation found in parks, streets, and private properties within urban areas. Green spaces include parks, gardens, green belts, and natural areas that offer recreational and aesthetic value. Both play a vital role in:

Mitigating Urban Heat Island Effect

Trees and vegetation cool the air through shade and evapotranspiration, helping to reduce temperatures in urban areas. This is particularly important for mitigating the urban heat island effect, where cities experience higher temperatures than their rural surroundings due to human activities and infrastructure.

Improving Air Quality

Urban forests and green spaces act as natural air filters, capturing airborne pollutants such as particulate matter, nitrogen oxides, and sulfur dioxide. This improves air quality and reduces respiratory and cardiovascular diseases among city residents.

Enhancing Biodiversity

Green spaces provide habitats for a wide range of species, supporting urban biodiversity. They serve as refuges for birds, insects, and small mammals, promoting ecological balance and enhancing the resilience of urban ecosystems.

Benefits of Urban Forests and Green Spaces

The integration of urban forests and green spaces into city planning offers numerous benefits:

Environmental Benefits

Environmental benefits include:

- Carbon Sequestration: Trees absorb carbon dioxide during photosynthesis, acting as carbon sinks and helping to mitigate climate change.
- Stormwater Management: Vegetation and permeable surfaces in green spaces absorb rainwater, reducing runoff and decreasing the risk of urban flooding. Green spaces also help recharge groundwater supplies.
- Erosion Control: Tree roots stabilize soil, preventing erosion and maintaining soil health.

Economic Benefits

Economic benefits include:

- Energy Savings: The cooling effect of trees reduces the need for air conditioning in buildings, leading to lower energy costs. Windbreaks provided by trees can also reduce heating costs in winter.
- Property Value Increase: Properties located near parks and green spaces typically have higher market values. Green spaces enhance the aesthetic appeal of neighborhoods, attracting residents and businesses.
- Tourism and Recreation: Well-maintained green spaces attract tourists and provide recreational opportunities for residents, contributing to the local economy.

Social Benefits

Social benefits include:

- Public Health: Access to green spaces encourages physical activity, reduces stress, and improves mental health. Green spaces provide venues for social interactions, fostering community cohesion.
- Aesthetic and Cultural Value: Green spaces enhance the visual appeal of urban areas, creating attractive and pleasant environments. They also serve as venues for cultural and community events.

Strategies for Implementing Urban Forests and Green Spaces

Effective implementation of urban forests and green spaces requires strategic planning, community involvement, and long-term maintenance. Key strategies include:

Urban Planning and Policy

Integrating green spaces into urban planning and zoning regulations ensures the preservation and creation of green areas. Policies should mandate minimum green space requirements for new developments and encourage the greening of existing urban areas.

Tree Planting and Maintenance Programs

Implementing tree planting initiatives, such as street tree programs and community tree planting events, increases urban tree canopy cover. Regular maintenance, including pruning, watering, and pest management, is essential for the health and longevity of urban forests.

Green Space Design

Designing multifunctional green spaces that accommodate diverse activities and user groups maximizes their utility and appeal. Green spaces should be accessible, safe, and equipped with amenities such as benches, playgrounds, and walking paths.

Community Engagement

Involving local communities in the planning, implementation, and stewardship of green spaces fosters a sense of ownership and responsibility. Community gardens, volunteer tree planting, and citizen science projects engage residents and encourage sustainable practices.

Partnerships and Collaboration

Collaborating with non-profit organizations, businesses, and government agencies leverages resources and expertise for green space projects. Public-private partnerships can fund and support the development and maintenance of urban green spaces.

Examples of Successful Urban Green Initiatives

Examples of successful urban green initiatives include the following.

Central Park, New York City, USA

Central Park is one of the most famous examples of urban green space, providing a vast oasis of greenery in the heart of Manhattan. The park offers recreational facilities, cultural events, and a habitat for wildlife, serving as a model for urban park design worldwide.

Singapore's Green Plan

Singapore has implemented a comprehensive green plan that integrates green spaces into its urban fabric. The city-state boasts an extensive network of parks, green corridors, and vertical greenery, enhancing its reputation as a "City in a Garden."

The High Line, New York City, USA

The High Line is an innovative urban park built on a disused elevated railway track. This linear park features diverse plantings, art installations, and seating areas, transforming a neglected space into a vibrant public amenity.

Seoul's Cheonggyecheon Stream

The restoration of the Cheonggyecheon Stream in Seoul transformed a buried waterway and highway into a 3.6-mile-long urban park. The project revitalized the city center, improved air quality, and provided a natural space for residents and tourists.

Sustainable Urban Drainage Systems

Sustainable Urban Drainage Systems (SUDS) are innovative approaches to managing stormwater in urban environments. Unlike traditional drainage systems, which rapidly channel water away through pipes and sewers, SUDS aim to mimic natural processes, promoting the infiltration, storage, and gradual release of rainwater. This approach helps mitigate flooding, improve water quality, enhance urban biodiversity, and create more resilient urban landscapes. This section explores the principles, benefits, and key components of SUDS, along with examples of successful implementation.

Principles of Sustainable Urban Drainage Systems

The core principles of SUDS focus on sustainability, water quality, and flood risk management. They aim to:

- Reduce Runoff: SUDS reduce the volume and rate of surface water runoff by promoting infiltration and storage. This helps prevent overwhelming drainage systems and reduces the risk of flooding.
- Improve Water Quality: By filtering and treating stormwater through natural processes, SUDS improve water quality

before it reaches watercourses. This reduces pollution and protects aquatic ecosystems.
- Enhance Amenity and Biodiversity: SUDS incorporate green spaces and vegetation, creating habitats for wildlife and enhancing the aesthetic and recreational value of urban areas.
- Integrate into the Urban Landscape: SUDS are designed to be aesthetically pleasing and multifunctional, blending seamlessly into the urban environment and providing additional benefits such as cooling and recreational spaces.

Benefits of Sustainable Urban Drainage Systems

Implementing SUDS offers numerous benefits, making cities more sustainable and resilient:

- Flood Risk Management: SUDS mitigate flood risks by slowing down and reducing the volume of stormwater runoff. This helps protect urban areas from flooding and minimizes damage to infrastructure and properties.
- Water Quality Improvement: SUDS filter pollutants from stormwater through natural processes, such as sedimentation, filtration, and biological uptake. This improves the quality of water entering rivers, lakes, and groundwater, benefiting both human and ecological health.
- Biodiversity and Habitat Creation: SUDS features, such as wetlands, ponds, and green roofs, provide habitats for various species of plants, birds, insects, and aquatic life. This enhances urban biodiversity and supports ecosystem services.
- Urban Cooling and Climate Resilience: Vegetated SUDS components, such as green roofs and rain gardens, help cool urban areas through evapotranspiration and shading. This reduces the urban heat island effect and improves the resilience of cities to climate change.
- Enhanced Aesthetics and Recreational Value: SUDS create attractive green spaces that can be used for recreation, relaxation, and community activities. These spaces enhance the quality of urban life and contribute to the well-being of residents.

Key Components of Sustainable Urban Drainage Systems

SUDS incorporate a variety of techniques and features that work together to manage stormwater sustainably. Key components include:

- Permeable Surfaces: Permeable pavements, such as porous asphalt, permeable concrete, and interlocking pavers, allow rainwater to infiltrate the ground rather than run off. These surfaces are used in parking lots, driveways, and walkways.
- Green Roofs: Green roofs are vegetated roof systems that absorb rainwater, reduce runoff, and provide insulation. They also enhance biodiversity and improve air quality.
- Rain Gardens and Bioretention Areas: Rain gardens and bioretention areas are landscaped depressions designed to collect, filter, and infiltrate stormwater. They are planted with native vegetation that can tolerate wet conditions and help remove pollutants.
- Swales and Filter Strips: Swales are shallow, vegetated channels that slow down and convey stormwater while allowing it to infiltrate. Filter strips are vegetated areas that treat runoff by filtering out sediments and pollutants.
- Ponds and Wetlands: Constructed ponds and wetlands store and treat large volumes of stormwater. They provide habitat for wildlife and recreational opportunities for people.
- Soakaways and Infiltration Trenches: Soakaways are underground structures that allow water to soak into the ground. Infiltration trenches are gravel-filled ditches that promote infiltration and groundwater recharge.

Examples of Successful SUDS Implementation

Examples of successful SUDS implementation include the following.

Queen Elizabeth Olympic Park, London, UK

The park features a range of SUDS components, including swales, rain gardens, and wetlands. These elements manage stormwater, enhance biodiversity, and provide recreational spaces for visitors.

Augustenborg Eco-City, Malmö, Sweden

This residential area incorporates green roofs, swales, ponds, and green spaces to manage stormwater sustainably. The project has successfully reduced flooding and created a vibrant urban environment.

Portland, Oregon, USA

Portland has implemented extensive SUDS infrastructure, including bioswales, green streets, and rain gardens. These features manage stormwater, improve water quality, and enhance the city's green spaces.

Benefits of Green Infrastructure in Urban Areas

Green infrastructure refers to the network of natural and semi-natural systems within urban environments that provide ecological, economic, and social benefits. Unlike traditional gray infrastructure, which primarily focuses on engineered solutions for urban challenges, green infrastructure utilizes nature-based approaches to enhance the resilience and sustainability of cities. This section outlines the myriad benefits of green infrastructure in urban areas, highlighting its role in promoting environmental health, economic vitality, and social well-being.

Environmental Benefits

Environmental benefits include:

- Stormwater Management: Green infrastructure effectively manages stormwater by mimicking natural hydrological

processes. Features such as green roofs, rain gardens, permeable pavements, and bioswales capture, filter, and infiltrate rainwater, reducing runoff and preventing urban flooding. This natural management of stormwater reduces the burden on municipal drainage systems and mitigates the risk of waterlogging and sewer overflows during heavy rainfall events.
- Improved Water Quality: By filtering pollutants from runoff, green infrastructure enhances the quality of water entering rivers, lakes, and groundwater. Vegetation and soil in green spaces trap sediments, heavy metals, and other contaminants, preventing them from polluting water bodies. This contributes to healthier aquatic ecosystems and safer water for human consumption and recreation.
- Urban Heat Island Mitigation: Vegetated areas cool the air through shading and evapotranspiration, reducing the urban heat island effect, where cities experience higher temperatures than surrounding rural areas. Green roofs, urban forests, and parks lower ambient temperatures, making urban environments more comfortable and reducing the need for air conditioning, which in turn lowers energy consumption and greenhouse gas emissions.
- Air Quality Improvement: Green infrastructure improves air quality by removing pollutants from the atmosphere. Trees and other vegetation capture particulate matter, nitrogen oxides, sulfur dioxide, and carbon dioxide, reducing pollution levels and improving respiratory health for urban residents. Increased greenery also contributes to the overall aesthetic and mental well-being of city dwellers.
- Biodiversity Enhancement: Green spaces provide habitats for a wide variety of plant and animal species, supporting urban biodiversity. Parks, green roofs, and urban gardens offer refuges for birds, insects, and small mammals, contributing to ecological balance and promoting a more resilient urban ecosystem. Biodiverse environments are also better able to withstand environmental stressors and recover from disturbances.

Economic Benefits

Economic benefits include:

- Energy Savings: Green infrastructure contributes to energy efficiency by insulating buildings and reducing the need for heating and cooling. Green roofs and walls act as natural insulators, maintaining indoor temperatures and reducing energy costs. Additionally, shaded streets and buildings lower the demand for air conditioning during hot weather, leading to significant energy savings.
- Increased Property Values: Properties located near green spaces typically have higher market values. The presence of parks, gardens, and tree-lined streets enhances the attractiveness of neighborhoods, making them more desirable places to live and work. This can lead to increased property tax revenues for municipalities and stimulate economic development.
- Job Creation: The planning, installation, and maintenance of green infrastructure create job opportunities in fields such as landscaping, horticulture, and environmental engineering. These jobs contribute to the local economy and provide employment for urban residents. Green infrastructure projects can also attract tourism and investment, further boosting economic activity.
- Cost Savings on Infrastructure: Green infrastructure can reduce the need for costly gray infrastructure upgrades by managing stormwater naturally. By reducing runoff and preventing flooding, green infrastructure extends the lifespan of existing drainage systems and minimizes maintenance costs. The multifunctional nature of green infrastructure also offers additional benefits, such as recreational spaces and improved urban aesthetics, at a lower cost compared to traditional infrastructure.

Social Benefits

Social benefits include:

- Improved Public Health: Access to green spaces promotes physical activity, reduces stress, and improves mental health. Parks, gardens, and recreational areas provide opportunities for exercise, social interaction, and relaxation, contributing to the overall well-being of urban residents. Green infrastructure also reduces exposure to air pollution and heat, leading to better respiratory and cardiovascular health.
- Community Cohesion: Green spaces serve as gathering places for community activities and events, fostering social cohesion and a sense of belonging. Community gardens and urban agriculture projects encourage residents to work together, strengthening social ties and promoting civic engagement. These spaces also provide educational opportunities for learning about nature, sustainability, and healthy living.
- Aesthetic and Cultural Value: Green infrastructure enhances the visual appeal of urban environments, creating pleasant and attractive landscapes. Beautiful, green spaces can improve the quality of life for residents, increase the appeal of urban areas for tourists, and contribute to a city's cultural and historical identity. Integrating green infrastructure into urban design reflects a commitment to sustainability and innovation, enhancing the city's reputation.

Chapter 5: Transportation and Mobility Solutions

Transportation and mobility are fundamental aspects of urban life, shaping the way people move, work, and live in cities. However, traditional transportation systems, heavily reliant on fossil fuels, contribute significantly to greenhouse gas emissions, air pollution, and traffic congestion. As urban populations grow and the impacts of climate change become more pronounced, there is an urgent need to rethink and transform urban mobility.

This chapter explores innovative transportation and mobility solutions that promote sustainability, reduce environmental impact, and enhance the quality of urban life. We will examine various strategies, including the integration of EVs, the expansion of public transit systems, the promotion of active transportation modes like cycling and walking, and the implementation of smart transportation technologies. Additionally, the chapter will highlight the role of urban planning and policy in creating sustainable transportation networks.

By embracing forward-thinking transportation strategies, cities can reduce their carbon footprint, improve air quality, and create more livable, efficient, and resilient urban environments. This chapter aims to provide a comprehensive understanding of the solutions available to transform urban mobility and the steps necessary to implement these solutions effectively.

Electric Vehicles and Charging Infrastructure

Electric vehicles (EVs) are transforming the transportation landscape, offering a cleaner and more sustainable alternative to traditional gasoline and diesel-powered vehicles. As urban areas strive to reduce greenhouse gas emissions, improve air quality, and enhance energy efficiency, the adoption of EVs plays a critical role in achieving these goals. However, the widespread deployment of

EVs requires the development of robust charging infrastructure to support the growing number of electric vehicles on the road. This section explores the benefits of EVs, the types of charging infrastructure, and the challenges and strategies for implementing EV charging networks in urban areas.

Benefits of Electric Vehicles

EVs offer numerous benefits, including reducing greenhouse gas emissions, improving air quality, and lowering operating costs, making them a crucial component of sustainable urban transportation.

Reduction in Greenhouse Gas Emissions

EVs produce zero tailpipe emissions, significantly reducing the amount of carbon dioxide (CO_2) and other greenhouse gases released into the atmosphere. When powered by renewable energy sources, EVs offer a nearly carbon-neutral mode of transportation.

Improved Air Quality

By eliminating exhaust emissions, EVs help reduce air pollutants such as nitrogen oxides (NOx) and particulate matter (PM), which are associated with respiratory and cardiovascular diseases. Improved air quality leads to better public health outcomes, particularly in densely populated urban areas.

Energy Efficiency

Electric motors are more efficient than internal combustion engines, converting a higher percentage of energy from the battery to power the vehicle. This efficiency reduces overall energy consumption and operating costs for vehicle owners.

Reduced Noise Pollution

EVs operate more quietly than conventional vehicles, contributing to lower noise pollution levels in urban environments. This can enhance the quality of life for city residents by creating a quieter and more pleasant living environment.

Economic Benefits

The growing EV market stimulates job creation in manufacturing, installation, and maintenance of EVs and charging infrastructure. Additionally, EV owners benefit from lower fuel and maintenance costs compared to traditional vehicles.

Types of EV Charging Infrastructure

EV charging infrastructure comes in various types, each designed to meet different needs and charging speeds.

Level 1 Charging

Level 1 charging uses a standard 120-volt household outlet and provides the slowest charging rate, typically adding about 2 to 5 miles of range per hour. This type of charging is suitable for overnight charging at home, where vehicles are parked for extended periods.

Level 2 Charging

Level 2 charging operates at 240 volts and can add 10 to 60 miles of range per hour, depending on the vehicle and charger specifications. These chargers are commonly installed in residential garages, workplaces, public parking facilities, and commercial locations. Level 2 charging provides a balance between charging speed and accessibility.

DC Fast Charging

Direct Current (DC) fast charging delivers high power levels, typically between 50 kW and 350 kW, allowing EVs to charge up to 80% of their battery capacity in 20 to 40 minutes. DC fast chargers are ideal for highway rest stops, urban fast-charging hubs, and locations where drivers need a quick recharge. These chargers are essential for long-distance travel and reducing charging downtime.

Wireless Charging

Wireless charging, or inductive charging, uses electromagnetic fields to transfer energy between a charging pad on the ground and a receiver on the vehicle. While still in the early stages of adoption, wireless charging offers a convenient and seamless charging experience, particularly for autonomous vehicles and public transit systems.

Challenges in Implementing EV Charging Infrastructure

Implementing EV charging infrastructure presents several challenges, including high initial costs, grid capacity issues, and ensuring accessibility.

Grid Capacity and Stability

The increasing number of EVs places additional demand on the electrical grid, requiring upgrades to grid infrastructure and capacity to ensure stability and reliability. Integrating renewable energy sources and energy storage solutions can help balance the load and support sustainable charging networks.

Installation and Maintenance Costs

The upfront costs of installing EV charging stations, particularly DC fast chargers, can be high. Financial incentives, grants, and public-private partnerships are essential to offset these costs and promote widespread deployment. Additionally, ongoing maintenance and operation costs must be considered.

Accessibility and Convenience

Ensuring that charging stations are conveniently located and accessible to all users, including those in multi-unit dwellings and underserved communities, is crucial. Public charging infrastructure must be strategically distributed to provide equitable access and support the diverse needs of urban residents.

Standardization and Interoperability

The lack of standardization and interoperability among different charging networks and connectors can create barriers for EV adoption. Developing universal standards and protocols for charging equipment and payment systems can enhance user experience and streamline the charging process.

Strategies for Expanding EV Charging Infrastructure

Strategies for expanding EV charging infrastructure focus on enhancing accessibility, promoting public-private partnerships, and integrating renewable energy sources.

Government Support and Policy

Governments at all levels play a critical role in promoting EV adoption and charging infrastructure development through policies, regulations, and incentives. This includes setting targets for EV deployment, offering tax credits and rebates for EV purchases and charger installations, and implementing building codes that mandate EV-ready infrastructure in new constructions and renovations.

Public-Private Partnerships

Collaborations between public agencies, private companies, and utility providers can leverage resources and expertise to expand charging networks. Public-private partnerships can facilitate the installation of charging stations in strategic locations, such as transit hubs, commercial centers, and public parking facilities.

Integration with Renewable Energy

Pairing EV charging stations with renewable energy sources, such as solar and wind power, can enhance the sustainability of EV charging networks. On-site renewable energy generation, combined with energy storage systems, can reduce the reliance on the grid and provide clean, reliable power for EV charging.

Innovative Business Models

Developing innovative business models, such as subscription-based charging services, pay-per-use schemes, and workplace charging programs, can increase the financial viability of charging infrastructure. Encouraging businesses to offer charging services as an amenity can also boost the adoption of EVs.

Community Engagement and Education

Raising awareness about the benefits of EVs and the availability of charging infrastructure is essential for driving adoption. Community outreach programs, educational campaigns, and incentives for early adopters can help build public support and accelerate the transition to electric mobility.

Public Transport Innovations

Public transportation systems are the backbone of urban mobility, providing efficient, cost-effective, and environmentally friendly alternatives to private car travel. As cities grow and face increasing challenges related to traffic congestion, air pollution, and climate change, innovative public transport solutions are essential for creating sustainable urban environments. This section explores various public transport innovations, including advancements in bus, rail, and shared mobility services, and highlights the benefits and implementation strategies for these innovations in urban areas.

Electrification of Public Transport

Electrification of public transport involves transitioning buses, trains, and other transit systems to electric power to reduce emissions and improve energy efficiency.

Electric Buses

Electric buses are rapidly becoming a cornerstone of sustainable urban transportation. These buses produce zero tailpipe emissions, reducing greenhouse gases and improving air quality. Modern electric buses offer quieter rides, lower operating costs, and reduced maintenance requirements compared to traditional diesel buses. Cities around the world are investing in electric bus fleets and charging infrastructure to support their deployment. Fast-charging stations at bus depots and along routes ensure that electric buses can operate throughout the day without significant downtime. Additionally, advancements in battery technology are extending the range and efficiency of electric buses.

Electrified Rail Systems

Electrification of rail systems, including trams, subways, and commuter trains, enhances the efficiency and sustainability of urban transit networks. Electrified rail systems offer high-capacity, reliable, and fast transportation options that reduce reliance on fossil fuels. Overhead wires, third rail systems, and on-board energy storage solutions are commonly used to power electrified trains. Regenerative braking systems capture and reuse energy, further improving the efficiency of electric rail networks.

Autonomous and Connected Public Transport

Autonomous and connected public transport leverages self-driving technology and smart communication systems to enhance the efficiency and safety of urban transit networks.

Autonomous Buses and Shuttles

Autonomous buses and shuttles are being tested and deployed in cities worldwide, offering the potential for safer, more efficient, and flexible public transport services. These self-driving vehicles use advanced sensors, cameras, and artificial intelligence to navigate urban environments and transport passengers without human intervention. Autonomous shuttles are particularly well-suited for last-mile connections, providing convenient links between major transit hubs and residential or commercial areas. By operating on fixed routes or demand-responsive services, autonomous shuttles can enhance the overall accessibility and convenience of public transport networks.

Connected Transport Systems

Connected transport systems use real-time data and communication technologies to improve the coordination and efficiency of public transport services. Vehicle-to-Infrastructure (V2I) and Vehicle-to-Vehicle (V2V) communication enable buses, trains, and traffic signals to interact seamlessly, reducing delays and optimizing traffic flow. Integrated ticketing and payment systems, powered by mobile apps and contactless payment technologies, streamline the passenger experience by allowing users to pay for multiple transport modes with a single transaction. Real-time information displays and mobile apps provide passengers with up-to-date information on schedules, delays, and service disruptions.

Sustainable Mobility Hubs

Sustainable mobility hubs integrate various modes of transportation, such as public transit, bike-sharing, and electric vehicle charging, to create seamless and eco-friendly travel options for urban commuters.

Multimodal Transport Hubs

Multimodal transport hubs integrate various modes of transportation, including buses, trains, trams, bike-sharing, and car-sharing services, into a single location. These hubs provide seamless connections between different transport modes, making it easier for passengers to switch from one mode to another. By consolidating transport services, multimodal hubs reduce the need for private car travel and encourage the use of sustainable transport options. Amenities such as waiting areas, retail outlets, and bike storage facilities enhance the convenience and attractiveness of these hubs.

Transit-Oriented Development (TOD)

Transit-oriented development (TOD) focuses on creating high-density, mixed-use communities centered around public transport stations. TOD encourages residents to live, work, and shop within walking distance of transit services, reducing the need for car travel and promoting sustainable urban growth. Effective TOD requires coordinated planning and investment in both transport infrastructure and urban development. Policies that support affordable housing, pedestrian-friendly streets, and green spaces are essential for creating vibrant, livable communities around transit hubs.

Flexible and On-Demand Transport Services

Flexible and on-demand transport services offer dynamic and responsive mobility solutions tailored to real-time passenger needs and preferences.

Microtransit

Microtransit services, often operated by private companies in partnership with public transit agencies, provide flexible, on-demand transport options for urban residents. Using mobile apps, passengers can request rides in real-time, and dynamically routed vehicles pick them up and drop them off at designated locations. Microtransit bridges the gap between traditional fixed-route services and ride-hailing, offering an efficient and cost-effective alternative for areas with lower demand or during off-peak hours. These services can complement existing public transport networks, enhancing coverage and convenience.

Demand-Responsive Transport (DRT)

Demand-responsive transport (DRT) systems adjust routes and schedules based on real-time passenger demand, optimizing vehicle utilization and reducing wait times. DRT can serve a wide range of urban transport needs, including paratransit services for individuals with disabilities, late-night transport, and suburban connections. Advanced algorithms and data analytics are used to plan and operate DRT services, ensuring that vehicles are deployed where and when they are needed most. By providing a more personalized and flexible transport experience, DRT can attract new users to public transport systems.

Integration of Active Transportation

Integration of active transportation promotes walking and cycling by incorporating dedicated infrastructure and connectivity into urban planning.

Bike-Sharing and E-Scooters

Bike-sharing and e-scooter programs offer convenient and sustainable transport options for short urban trips. These services reduce traffic congestion, lower emissions, and promote physical activity among city residents. Dockless systems, which allow users to pick up and drop off bikes or scooters at any location within a designated area, have increased the flexibility and accessibility of these services. Integrating bike-sharing and e-scooter programs with public transport networks further enhances urban mobility.

Pedestrian Infrastructure

Investing in pedestrian infrastructure, such as sidewalks, crosswalks, and pedestrian plazas, encourages walking as a mode of transport. Safe and attractive pedestrian environments support healthy, active lifestyles and reduce the reliance on motorized transport. Urban planning initiatives that prioritize walkability, such as pedestrian zones, traffic calming measures, and greenways, create more livable and sustainable cities.

Challenges and Implementation Strategies

There are various actionable plans necessary for successfully adopting new transportation innovations and infrastructure, including the following.

Funding and Investment

Securing funding for public transport innovations can be challenging. Public-private partnerships, government grants, and innovative financing mechanisms are essential for supporting the development and deployment of new transport technologies and infrastructure.

Policy and Regulation

Effective policies and regulations are crucial for promoting sustainable transport innovations. Governments must establish supportive frameworks that encourage investment in public transport, incentivize the use of clean technologies, and ensure the safety and reliability of new transport modes.

Public Acceptance and Engagement

Building public support for transport innovations requires transparent communication, community engagement, and education. Demonstrating the benefits of new transport solutions and addressing concerns through pilot projects and stakeholder consultations can foster acceptance and adoption.

Bike-sharing and Micro-mobility

As urban areas continue to expand and grapple with traffic congestion, air pollution, and the need for sustainable transportation solutions, bike-sharing and micro-mobility have emerged as effective alternatives to traditional car travel. These innovative mobility options provide flexible, cost-effective, and environmentally friendly means of transportation for short urban trips. This section delves into the concepts of bike-sharing and micro-mobility, explores their benefits, discusses implementation strategies, and addresses the challenges associated with their adoption.

Benefits of Bike-sharing and Micro-mobility

There are multiple benefits of bike-sharing and micro-mobility, including the following:

- Environmental Benefits:
    - Reduction in Emissions: By offering a zero-emission alternative to car travel, bike-sharing and micro-mobility help reduce greenhouse gas emissions and air pollutants. This contributes to cleaner air and mitigates the impact of climate change.
    - Decreased Traffic Congestion: By providing an alternative to driving, these services can alleviate traffic congestion in urban areas, leading to smoother traffic flow and shorter travel times.
- Health and Wellness:
    - Physical Activity: Bike-sharing encourages physical activity, promoting cardiovascular health and reducing the risk of lifestyle-related diseases. Regular use of bicycles or e-bikes can help urban residents integrate exercise into their daily routines.
    - Mental Health Benefits: Active transportation modes, such as cycling, are associated with improved mental

health, reduced stress levels, and enhanced overall well-being.
- Economic Advantages:
    o Cost Savings: Bike-sharing and micro-mobility services are often more affordable than owning and maintaining a private vehicle. They provide a cost-effective transportation option for short trips, reducing the financial burden on users.
    o Job Creation: The deployment and maintenance of bike-sharing and micro-mobility systems create employment opportunities in areas such as fleet management, customer service, and vehicle maintenance.
- Urban Mobility and Accessibility:
    o First and Last Mile Connectivity: These services bridge the gap between public transportation stops and final destinations, improving the accessibility and convenience of transit systems. This enhances overall mobility and makes public transport more attractive.
    o Flexible and Convenient: The ability to pick up and drop off vehicles at various locations provides users with flexibility and convenience, making it easier to navigate urban areas efficiently.

Implementation Strategies for Bike-sharing and Micro-mobility

Implementation strategies for bike-sharing and micro-mobility focus on infrastructure development, regulatory support, and community engagement to ensure effective adoption and operation.

Infrastructure Development

Dedicated Lanes and Parking: Developing dedicated bike lanes and secure parking facilities for bicycles and e-scooters is essential for the safety and convenience of users. Adequate infrastructure encourages more people to adopt these modes of transport. Charging Stations for E-bikes and E-scooters: Installing charging stations at strategic locations ensures that electric vehicles remain operational and accessible. This infrastructure supports the widespread use of e-bikes and e-scooters.

Integration with Public Transport

Seamless Connectivity: Integrating bike-sharing and micro-mobility services with public transport systems creates a cohesive urban mobility network. Users can transition smoothly between different modes of transport, enhancing overall efficiency. Unified Payment Systems: Implementing unified payment systems that allow users to pay for multiple transportation services with a single app or card simplifies the user experience and promotes the use of multimodal transport options.

Policy and Regulation

Supportive Policies: Governments should establish policies that support the deployment and operation of bike-sharing and micro-mobility services. This includes setting standards for safety, vehicle quality, and service reliability. Regulatory Frameworks: Clear regulatory frameworks are necessary to address issues such as parking, speed limits, and rider behavior. Effective regulation ensures the orderly operation of these services and minimizes potential disruptions.

Public Awareness and Engagement:

Education Campaigns: Public awareness campaigns can educate residents about the benefits of bike-sharing and micro-mobility, as well as safe riding practices. Promoting these services through various channels encourages wider adoption. Community Involvement: Engaging local communities in the planning and implementation of these services fosters a sense of ownership and ensures that the systems meet the needs of residents.

Challenges and Considerations

Some of the challenges and considerations include:

- Safety Concerns: Ensuring the safety of riders is paramount. Cities must invest in infrastructure such as protected bike lanes and proper lighting. Additionally, educating riders on safe practices and enforcing traffic regulations can reduce the risk of accidents.
- Maintenance and Vandalism: Regular maintenance of vehicles and addressing issues of vandalism and theft are critical for the sustainability of bike-sharing and micro-mobility services. Service providers must implement robust maintenance protocols and security measures.
- Equity and Accessibility: Ensuring that bike-sharing and micro-mobility services are accessible to all urban residents, including low-income communities and those with disabilities, is essential. Strategies such as subsidized pricing and adaptive vehicles can promote inclusivity.
- Environmental Impact of Manufacturing and Disposal: While these services reduce emissions from transportation, the environmental impact of manufacturing and disposing of vehicles must be considered. Sustainable practices in production, maintenance, and end-of-life disposal are necessary to minimize environmental harm.

Reducing Urban Traffic Congestion

Urban traffic congestion is a persistent issue affecting cities worldwide, leading to increased travel times, air pollution, and economic losses. As urban populations continue to grow, finding effective solutions to reduce traffic congestion becomes increasingly critical. This section explores various strategies for alleviating traffic congestion in urban areas, including improvements in public transportation, the promotion of active transportation modes, smart traffic management systems, and policy measures aimed at reducing car dependency.

Enhancing Public Transportation

Enhancing public transportation involves improving service efficiency, expanding coverage, and integrating advanced technologies to better meet urban mobility needs.

Expanding and Improving Public Transit Networks

Investing in the expansion and enhancement of public transportation networks is crucial for reducing traffic congestion. High-capacity transit options such as subways, light rail, and bus rapid transit (BRT) systems can move large numbers of passengers efficiently, reducing the number of private vehicles on the road. Improving the reliability, frequency, and coverage of public transit services makes them a more attractive option for commuters. Ensuring that transit systems are accessible, affordable, and convenient can encourage more people to use public transportation instead of driving.

Integrated Multimodal Transport

Developing integrated multimodal transport systems that facilitate seamless transitions between different modes of transportation can enhance the overall efficiency of urban mobility. This includes connecting public transit with bike-sharing, e-scooter services, and pedestrian pathways.Implementing unified ticketing systems that allow passengers to use a single payment method for multiple transport modes simplifies the user experience and promotes the use of public transit and active transportation.

Promoting Active Transportation

Promoting active transportation encourages walking and cycling through the development of safe and accessible infrastructure.

Developing Pedestrian and Cycling Infrastructure

Creating safe and attractive infrastructure for walking and cycling is essential for promoting active transportation. This includes building dedicated bike lanes, pedestrian pathways, and safe crossings, as well as providing amenities such as bike parking and repair stations. Designing urban spaces that prioritize pedestrians and cyclists over motor vehicles can encourage more people to choose active transportation for short trips, reducing traffic congestion and improving public health.

Encouraging the Use of E-Bikes and E-Scooters

E-bikes and e-scooters offer a convenient and sustainable alternative for short urban trips. Providing charging stations and designated parking areas for these vehicles can support their adoption and integration into the urban mobility network. Implementing regulations and safety measures to ensure the responsible use of e-bikes and e-scooters can enhance their appeal and reduce potential conflicts with other road users.

Implementing Smart Traffic Management Systems

Implementing smart traffic management systems leverages technology to optimize traffic flow and reduce congestion in urban areas.

Real-Time Traffic Monitoring and Management

Utilizing advanced technologies such as sensors, cameras, and GPS to monitor traffic conditions in real time can enable more efficient traffic management. Traffic management centers can use this data to adjust traffic signals, manage congestion, and respond to incidents more effectively. Dynamic traffic signal control systems that adapt to real-time traffic conditions can optimize traffic flow and reduce delays at intersections. These systems can prioritize public transit vehicles and emergency services to improve overall traffic efficiency.

Intelligent Transportation Systems (ITS)

ITS technologies, including connected vehicles and smart infrastructure, can enhance communication between vehicles and traffic management systems. This can lead to improved traffic flow, reduced congestion, and enhanced safety. Implementing smart parking systems that provide real-time information on parking availability can reduce the time drivers spend searching for parking, thereby reducing congestion and emissions.

Policy Measures to Reduce Car Dependency

Policy measures to reduce car dependency focus on encouraging the use of public transportation, cycling, and walking through incentives and regulations.

Congestion Pricing

Implementing congestion pricing schemes that charge drivers for entering high-traffic areas during peak times can discourage the use of private vehicles and reduce traffic congestion. Revenue generated from congestion pricing can be reinvested in public transportation and infrastructure improvements. Variable pricing based on time of day and traffic conditions can incentivize drivers to travel during off-peak hours, further alleviating congestion.

Carpooling and Ride-Sharing Programs

Encouraging carpooling and ride-sharing can reduce the number of vehicles on the road, especially during peak hours. Creating dedicated carpool lanes and providing incentives for carpooling can enhance participation in these programs. Partnering with ride-sharing companies to integrate their services with public transit can provide flexible transportation options and reduce the need for private car ownership.

Urban Planning and Land Use Policies

Adopting urban planning and land use policies that promote compact, mixed-use development can reduce the need for long-distance travel and support the use of public transportation, walking, and cycling. Creating transit-oriented developments (TODs) that concentrate housing, jobs, and amenities around public transit hubs can enhance accessibility and reduce car dependency.

Chapter 6: Building Smart and Sustainable Cities

As the global population increasingly gravitates towards urban centers, the need for smart and sustainable city planning has never been more critical. Cities are the hubs of economic activity, innovation, and cultural exchange, but they also face significant challenges, including environmental degradation, resource depletion, and social inequality. Building smart and sustainable cities involves integrating advanced technologies with sustainable practices to create urban environments that are resilient, efficient, and equitable.

This chapter explores the core principles and strategies for developing smart and sustainable cities. We will delve into the use of information and communication technologies (ICT) to enhance urban infrastructure, improve resource management, and foster community engagement. Additionally, the chapter will highlight the importance of sustainable urban planning, which prioritizes green spaces, efficient transportation systems, and energy-efficient buildings. Through a comprehensive examination of these elements, we aim to provide a roadmap for cities to transition towards a smarter and more sustainable future.

By embracing innovation and sustainability, cities can address the challenges of rapid urbanization and climate change, ensuring a high quality of life for their residents while minimizing their environmental footprint. This chapter will outline practical steps and best practices for city planners, policymakers, and stakeholders to collaboratively build the cities of tomorrow.

Energy-efficient Building Designs

Energy-efficient building designs are essential for creating sustainable urban environments. As cities grow and the demand for energy increases, it is crucial to adopt design strategies that reduce energy consumption, lower greenhouse gas emissions, and promote environmental stewardship. This section explores the principles of energy-efficient building designs, the various techniques and technologies employed, and the benefits of implementing these designs in urban settings.

Principles of Energy-efficient Building Design

Principles of energy-efficient building design emphasize optimizing energy use through smart architecture, sustainable materials, and advanced technologies:

- Passive Design Strategies:
    o Orientation and Layout: The orientation of a building significantly impacts its energy efficiency. Buildings should be positioned to maximize natural light and solar heat gain during winter while minimizing heat gain during summer. Strategic placement of windows, skylights, and other openings can enhance natural ventilation and daylighting.
    o Insulation and Thermal Mass: Proper insulation in walls, roofs, and floors reduces heat transfer, maintaining a comfortable indoor temperature with minimal energy use. Thermal mass materials, such as concrete and brick, absorb and store heat during the day and release it at night, helping to regulate indoor temperatures naturally.
    o Shading and Glazing: Shading devices, such as overhangs, louvers, and pergolas, can block direct sunlight and reduce cooling loads. High-performance glazing with low-emissivity coatings can minimize heat gain while allowing natural light to enter the building.
- Active Design Strategies:
    o Efficient HVAC Systems: Heating, ventilation, and air conditioning (HVAC) systems account for a significant portion of a building's energy consumption. High-efficiency HVAC systems, combined with smart thermostats and zoning controls,

optimize energy use while maintaining indoor comfort.
- o Lighting Systems: Energy-efficient lighting, such as LED bulbs and smart lighting controls, can significantly reduce electricity consumption. Daylighting strategies that maximize natural light can also minimize the need for artificial lighting during the day.
- o Renewable Energy Integration: Incorporating renewable energy sources, such as solar panels and wind turbines, into building designs can offset energy consumption and reduce reliance on fossil fuels. Buildings can generate their own electricity, contributing to a more sustainable energy grid.

- Water Conservation:
    - o Low-flow Fixtures: Installing low-flow faucets, showerheads, and toilets can reduce water consumption without compromising performance. Water-efficient appliances, such as dishwashers and washing machines, also contribute to water conservation.
    - o Rainwater Harvesting and Greywater Recycling: Systems that capture and reuse rainwater and greywater for irrigation, flushing toilets, and other non-potable uses can significantly reduce the demand for fresh water.
- Materials and Construction:
    - o Sustainable Materials: Using sustainable, locally sourced, and recycled materials reduces the environmental impact of construction. Materials with low embodied energy, such as bamboo, reclaimed wood, and recycled metal, are preferred for their minimal carbon footprint.
    - o Green Roofs and Walls: Green roofs and living walls provide insulation, reduce heat island effects, and improve air quality. They also create habitats for urban wildlife and contribute to stormwater management.

Techniques and Technologies for Energy-efficient Building Designs

Techniques and technologies for energy-efficient building designs include the following.

1. Building Information Modeling (BIM):

- BIM is a digital representation of the physical and functional characteristics of a building. It allows architects and engineers to simulate energy performance, optimize designs, and identify potential issues before construction begins. BIM facilitates collaboration among stakeholders, ensuring that energy efficiency is considered at every stage of the project.

2. Energy Modeling and Simulation:

- Energy modeling software predicts a building's energy consumption based on its design, location, and usage patterns. This tool helps designers evaluate the impact of various design choices on energy efficiency, enabling them to make informed decisions that enhance performance.

3. Smart Building Technologies:

- Smart building technologies, such as sensors, automation systems, and IoT devices, monitor and control energy use in real-time. These technologies can adjust lighting, heating, and cooling based on occupancy and environmental conditions, optimizing energy consumption and reducing waste.

4. Prefabrication and Modular Construction:

- Prefabrication involves constructing building components off-site in a controlled environment. This method reduces material waste, improves construction quality, and shortens project timelines.

Modular construction, a form of prefabrication, allows for flexible and scalable building designs that can adapt to changing needs.

Benefits of Energy-efficient Building Designs

Benefits of energy-efficient building designs include reduced energy consumption, lower utility costs, and enhanced occupant comfort.

1. Environmental Benefits:

- Reduced Carbon Emissions: Energy-efficient buildings consume less energy, leading to lower greenhouse gas emissions. This contributes to global efforts to combat climate change and improve air quality.

- Resource Conservation: By using less energy and water, energy-efficient buildings conserve natural resources and reduce the strain on local infrastructure and ecosystems.

2. Economic Benefits:

- Lower Operating Costs: Energy-efficient buildings have lower utility bills due to reduced energy and water consumption. Over time, these savings can offset the initial investment in energy-efficient technologies and design strategies.

- Increased Property Value: Buildings with energy-efficient features often have higher market values and attract environmentally conscious buyers and tenants. Certification programs, such as LEED and BREEAM, can enhance a building's reputation and marketability.

3. Social Benefits:

- Improved Comfort and Health: Energy-efficient buildings provide a more comfortable indoor environment with better temperature control, air quality, and natural light. This can enhance the well-being and productivity of occupants.

- Community Resilience: Energy-efficient buildings contribute to the overall resilience of urban areas by reducing the demand on energy grids and water systems. They can also serve as examples of sustainable living, inspiring other developments and fostering a culture of sustainability.

Smart Building Technologies

Smart building technologies are revolutionizing the way buildings operate, making them more efficient, comfortable, and sustainable. By integrating advanced systems and devices, these technologies enable real-time monitoring, control, and optimization of building operations. This section explores the various smart building technologies, their benefits, and the strategies for implementing them effectively in urban environments.

Overview of Smart Building Technologies

Smart building technologies encompass a range of systems and devices that enhance the functionality, efficiency, and sustainability of buildings. These technologies use sensors, automation systems, and data analytics to collect and analyze information about building operations, allowing for more informed decision-making and improved performance. Key components of smart building technologies include:

1. Building Management Systems (BMS):

- BMS are centralized platforms that control and monitor a building's mechanical and electrical systems, such as HVAC, lighting, and security. These systems use sensors and automation to optimize energy use, enhance comfort, and ensure safety.

2. Internet of Things (IoT):

- IoT devices, such as smart thermostats, lighting controls, and occupancy sensors, collect data and communicate with each other to create a connected building environment. These devices enable real-time monitoring and control of building systems, improving efficiency and responsiveness.

3. Energy Management Systems (EMS):

- EMS are designed to monitor, control, and optimize energy consumption in buildings. These systems use data from various sensors and meters to analyze energy use patterns and implement strategies to reduce consumption and costs.

4. Smart Lighting Systems:

- Smart lighting systems use sensors and automation to adjust lighting levels based on occupancy, daylight availability, and user preferences. These systems can reduce energy consumption and enhance the comfort and productivity of occupants.

5. HVAC Optimization:

- Advanced HVAC systems use sensors and data analytics to optimize heating, ventilation, and air conditioning operations. These systems can adjust temperature and airflow based on occupancy, weather conditions, and energy prices, improving energy efficiency and indoor comfort.

6. Security and Access Control:

- Smart security systems use cameras, sensors, and access control devices to monitor and manage building security. These systems can detect and respond to security threats in real-time, enhancing safety and protection.

Benefits of Smart Building Technologies

The adoption of smart building technologies offers numerous benefits, including:

1. Energy Efficiency:

- Smart building technologies optimize energy use by adjusting systems based on real-time data and conditions. This can significantly reduce energy consumption and costs, contributing to sustainability and environmental goals.

2. Enhanced Comfort and Productivity:

- By maintaining optimal indoor conditions, smart building technologies improve occupant comfort and well-being. Better lighting, temperature control, and air quality can enhance productivity and satisfaction among building users.

3. Operational Efficiency:

- Automation and data analytics streamline building operations, reducing the need for manual intervention and maintenance. This can lead to cost savings and improved reliability of building systems.

4. Safety and Security:

- Advanced security systems protect buildings and their occupants by detecting and responding to threats in real-time. Smart access control systems ensure that only authorized individuals can enter specific areas, enhancing overall security.

5. Sustainability:

- Smart building technologies support sustainable practices by reducing energy and water consumption, minimizing waste, and lowering greenhouse gas emissions. These technologies help buildings achieve sustainability certifications and comply with environmental regulations.

Implementing Smart Building Technologies

Effective implementation of smart building technologies requires careful planning, investment, and collaboration among stakeholders. Key strategies for successful deployment include:

1. Assessment and Planning:

- Conducting a thorough assessment of the building's current systems and identifying areas for improvement is the first step. This involves evaluating energy use, operational efficiency, and occupant needs. Developing a comprehensive plan that outlines the goals, budget, and timeline for implementation is essential.

2. Choosing the Right Technologies:

- Selecting the appropriate smart building technologies based on the building's specific needs and goals is crucial. This includes choosing reliable and scalable systems that can integrate with existing infrastructure and future upgrades.

3. Integration and Interoperability:

- Ensuring that different smart building systems can communicate and work together seamlessly is vital for maximizing their benefits. This requires selecting technologies that support open standards and protocols, facilitating interoperability and integration.

4. Data Management and Analytics:

- Effective data management is critical for the success of smart building technologies. Implementing robust data collection, storage, and analysis systems enables building managers to make informed decisions and optimize operations.

5. Training and Support:

- Providing training and support for building staff and occupants is essential for the successful adoption of smart building technologies. This includes educating users on how to use the systems effectively and addressing any concerns or issues that may arise.

6. Monitoring and Maintenance:

- Regular monitoring and maintenance of smart building systems ensure their continued performance and reliability. This involves routine inspections, software updates, and addressing any technical issues promptly.

Challenges and Considerations

While smart building technologies offer significant benefits, there are several challenges and considerations to keep in mind:

1. Initial Costs:

- The upfront costs of implementing smart building technologies can be substantial. However, the long-term savings in energy and operational costs can offset these initial investments.

2. Data Security and Privacy:

- The increased connectivity and data collection associated with smart building technologies raise concerns about data security and privacy. Implementing robust cybersecurity measures and ensuring compliance with data protection regulations are essential.

3. Scalability:

- Ensuring that smart building technologies can scale with the building's needs and future upgrades is crucial. Selecting flexible and scalable systems can accommodate growth and technological advancements.

4. User Acceptance:

- Gaining acceptance and buy-in from building occupants and staff is essential for the successful implementation of smart building technologies. Addressing user concerns and providing adequate training can facilitate a smooth transition.

Retrofitting Existing Buildings

Retrofitting existing buildings is a crucial strategy for improving energy efficiency, reducing carbon emissions, and enhancing the sustainability of urban environments. As a significant portion of the building stock in cities is comprised of older structures that were not designed with modern energy standards in mind, retrofitting offers a cost-effective and practical approach to upgrading these buildings. This section explores the principles, techniques, benefits, and challenges of retrofitting existing buildings for energy efficiency and sustainability.

Principles of Retrofitting Existing Buildings

Retrofitting involves upgrading the building's systems and components to improve its energy performance and environmental impact. The primary goals of retrofitting are to:

1. Enhance Energy Efficiency:

- Implementing measures that reduce energy consumption, such as improving insulation, upgrading HVAC systems, and installing energy-efficient lighting.

2. Reduce Carbon Emissions:

- Lowering the building's carbon footprint by utilizing renewable energy sources and improving energy management practices.

3. Improve Comfort and Health:

- Enhancing indoor environmental quality through better air circulation, temperature control, and lighting, contributing to the well-being of occupants.

4. Extend Building Lifespan:

- Upgrading structural and functional elements to prolong the building's usability and value.

Techniques for Retrofitting Existing Buildings

1. Building Envelope Improvements:

- Insulation: Adding or upgrading insulation in walls, roofs, and floors reduces heat transfer, maintaining comfortable indoor temperatures with less energy use.

- Windows and Doors: Replacing single-pane windows with double or triple-pane options and improving door seals minimize heat loss and drafts.

- Air Sealing: Sealing gaps and cracks in the building envelope prevents air leakage, enhancing the efficiency of heating and cooling systems.

2. HVAC System Upgrades:

- Efficient Heating and Cooling: Replacing outdated HVAC systems with high-efficiency models, such as heat pumps, reduces energy consumption and improves indoor comfort.

- Smart Thermostats: Installing programmable thermostats and building management systems allows for precise control of heating and cooling, optimizing energy use.

- Ventilation: Upgrading ventilation systems to ensure adequate air exchange while minimizing energy loss improves indoor air quality and energy efficiency.

3. Lighting and Electrical Systems:

- LED Lighting: Replacing incandescent and fluorescent bulbs with LED lighting significantly reduces energy consumption and maintenance costs.

- Lighting Controls: Implementing occupancy sensors, daylight sensors, and dimmers adjusts lighting levels based on occupancy and natural light availability, further reducing energy use.

- Energy-Efficient Appliances: Upgrading to ENERGY STAR-rated appliances and equipment reduces electricity consumption and operating costs.

4. Renewable Energy Integration:

- Solar Panels: Installing photovoltaic (PV) panels on roofs or facades generates clean, renewable electricity, reducing dependence on grid power.

- Wind Turbines: Small wind turbines can be used in suitable locations to generate additional renewable energy.

- Geothermal Systems: Geothermal heat pumps use the stable temperature of the ground to provide efficient heating and cooling.

5. Water Efficiency Measures:

- Low-Flow Fixtures: Installing low-flow faucets, showerheads, and toilets reduces water consumption without sacrificing performance.

- Greywater Recycling: Systems that collect and reuse greywater for non-potable uses, such as irrigation and flushing toilets, decrease water demand.

- Rainwater Harvesting: Capturing and storing rainwater for irrigation and other uses reduces reliance on municipal water supplies.

Benefits of Retrofitting Existing Buildings

1. Environmental Benefits:

- Reduced Carbon Footprint: By enhancing energy efficiency and incorporating renewable energy sources, retrofitting significantly lowers the building's carbon emissions.

- Resource Conservation: Retrofitting conserves resources by reducing energy and water consumption, contributing to environmental sustainability.

2. Economic Benefits:

- Energy Cost Savings: Improved energy efficiency results in lower utility bills, providing substantial cost savings over time.

- Increased Property Value: Energy-efficient buildings are more attractive to buyers and tenants, often commanding higher market values and rental rates.

- Incentives and Rebates: Many governments and utility companies offer financial incentives, rebates, and tax credits for energy-efficient upgrades, offsetting initial investment costs.

3. Social Benefits:

- Improved Comfort and Health: Enhancing indoor environmental quality through better insulation, ventilation, and lighting improves occupant comfort and health.

- Job Creation: Retrofitting projects create jobs in construction, engineering, and related fields, boosting local economies.

Challenges of Retrofitting Existing Buildings

1. Initial Costs:

- The upfront costs of retrofitting can be significant, particularly for extensive upgrades. However, the long-term savings and benefits often outweigh these initial expenses.

2. Technical Complexity:

- Retrofitting older buildings can be technically challenging, requiring careful planning and expertise to address structural limitations and integrate new systems seamlessly.

3. Disruption to Occupants:

- Retrofit projects can disrupt building occupants, necessitating careful scheduling and communication to minimize inconvenience.

4. Regulatory Hurdles:

- Navigating building codes, zoning laws, and historical preservation regulations can complicate retrofitting projects, requiring coordination with local authorities.

Policies and Incentives for Sustainable Construction

Sustainable construction is essential for reducing the environmental impact of buildings and promoting a greener future. Governments and organizations worldwide are increasingly recognizing the importance of implementing policies and incentives to encourage sustainable construction practices. These measures aim to reduce greenhouse gas emissions, improve energy efficiency, and promote the use of sustainable materials. This section explores various policies and incentives that support sustainable construction, their benefits, and the challenges associated with their implementation.

Policies Supporting Sustainable Construction

1. Building Codes and Standards:

- Governments implement building codes and standards that mandate energy efficiency and sustainable practices in construction. These regulations often include requirements for insulation, ventilation, lighting, and the use of renewable energy sources.

- Examples include the International Green Construction Code (IgCC) and the Leadership in Energy and Environmental Design (LEED) certification system. These codes and standards provide a framework for designing and constructing buildings that meet specific sustainability criteria.

2. Energy Efficiency Regulations:

- Energy efficiency regulations set minimum performance standards for buildings and their systems. These regulations aim to reduce energy consumption and encourage the adoption of energy-efficient technologies.

- Policies such as the European Union's Energy Performance of Buildings Directive (EPBD) require buildings to meet energy performance standards and undergo regular energy audits to ensure compliance.

3. Renewable Energy Mandates:

- Renewable energy mandates require a certain percentage of a building's energy to come from renewable sources such as solar, wind, or geothermal energy. These mandates help reduce reliance on fossil fuels and promote the integration of clean energy technologies.

- For instance, California's Title 24 Building Energy Efficiency Standards mandate that new residential buildings incorporate solar photovoltaic systems.

4. Sustainable Material Requirements:

- Policies that encourage or require the use of sustainable materials in construction help reduce the environmental impact of building projects. This includes promoting the use of recycled, locally sourced, and low-emission materials.

- Green procurement policies, such as those implemented by various municipalities, prioritize the use of sustainable materials in public construction projects.

5. Waste Management Regulations:

- Construction waste management regulations aim to minimize waste generation and promote recycling and reuse of materials. These policies help reduce the environmental impact of construction activities and support the circular economy.

- Examples include regulations that require construction projects to develop and implement waste management plans, as seen in the UK's Site Waste Management Plans Regulations.

Incentives for Sustainable Construction

1. Financial Incentives:

- Governments offer financial incentives such as grants, rebates, tax credits, and low-interest loans to encourage sustainable construction practices. These incentives help offset the initial costs of implementing energy-efficient technologies and sustainable materials.

- For example, the U.S. Federal Energy Tax Credit provides tax credits for residential renewable energy installations, while various state and local programs offer rebates for energy-efficient upgrades.

2. Green Building Certification Programs:

- Certification programs such as LEED, BREEAM (Building Research Establishment Environmental Assessment Method), and ENERGY STAR recognize and reward buildings that meet specific sustainability criteria. Certified buildings often qualify for financial incentives and can command higher market values.

- These programs provide a standardized framework for evaluating and certifying the sustainability performance of buildings, encouraging developers to adopt green building practices.

3. Zoning and Land Use Incentives:

- Zoning and land use policies can include incentives for sustainable construction, such as density bonuses, expedited permitting, and reduced fees for projects that meet sustainability criteria.

- Density bonuses allow developers to build at higher densities in exchange for incorporating green building features, promoting sustainable urban development.

4. Public-Private Partnerships:

- Public-private partnerships (PPPs) can facilitate sustainable construction by leveraging the resources and expertise of both sectors. These partnerships can fund and implement large-scale sustainable projects that might be challenging for the public or private sector to undertake alone.

- Examples include the development of green infrastructure, energy-efficient public buildings, and renewable energy installations.

Benefits of Policies and Incentives

1. Environmental Benefits:

- Policies and incentives for sustainable construction help reduce greenhouse gas emissions, conserve natural resources, and minimize waste. By promoting energy efficiency and the use of renewable energy, these measures contribute to mitigating climate change and protecting the environment.

2. Economic Benefits:

- Sustainable construction practices can lead to significant cost savings over the lifecycle of a building. Energy-efficient buildings have lower operating costs, and incentives can reduce the upfront costs of sustainable technologies and materials.

- Additionally, green buildings often have higher property values and can attract tenants willing to pay a premium for environmentally friendly spaces.

3. Social Benefits:

- Sustainable buildings provide healthier and more comfortable living and working environments. Improved indoor air quality, natural lighting, and temperature control contribute to the well-being and productivity of occupants.

- Policies and incentives that promote sustainable construction can also create jobs in the green building sector, contributing to economic growth and workforce development.

Challenges of Implementing Policies and Incentives

1. Initial Costs:

- The upfront costs of sustainable construction can be higher than traditional building methods, posing a barrier for developers and property owners. Financial incentives and supportive policies are essential to offset these initial expenses and encourage adoption.

2. Regulatory Complexity:

- Navigating the various regulations, standards, and certification programs can be complex and time-consuming. Simplifying the regulatory framework and providing clear guidance can help stakeholders understand and comply with sustainable construction requirements.

3. Market Demand:

- While the demand for green buildings is growing, it is not yet universal. Educating consumers and promoting the benefits of sustainable construction can drive market demand and encourage wider adoption.

Chapter 7: Data-Driven Urban Planning

In an era of rapid urban growth and increasing complexity, data-driven urban planning has emerged as a critical tool for designing, managing, and optimizing cities. By leveraging big data, advanced analytics, and smart technologies, urban planners can gain deep insights into city dynamics, resource use, and resident behavior.

This chapter explores the potential of data-driven urban planning to enhance decision-making, efficiency, and sustainability. We will discuss the tools and methodologies for collecting, analyzing, and interpreting urban data, along with the benefits and challenges of these approaches. Practical applications, such as optimizing public transportation, energy management, and improving public health and safety, will be highlighted.

Data-driven urban planning enables cities to become more responsive, resilient, and inclusive. This chapter aims to provide a comprehensive overview of how data can address urban challenges and promote sustainable development, offering valuable insights for policymakers, planners, and stakeholders dedicated to building smarter cities for the future.

Role of Big Data and IoT in Urban Planning

The integration of Big Data and the Internet of Things (IoT) into urban planning marks a significant advancement in how cities are designed, managed, and optimized. By leveraging vast amounts of data and interconnected devices, urban planners can make more informed decisions that enhance the efficiency, sustainability, and livability of urban environments. This section explores the role of Big Data and IoT in urban planning, detailing their applications, benefits, and the challenges they present.

Understanding Big Data and IoT

Understanding Big Data and IoT involves recognizing how vast data sets and interconnected devices can drive smarter urban planning and management.

1. Big Data:

- Big Data refers to extremely large datasets that can be analyzed computationally to reveal patterns, trends, and associations, especially relating to human behavior and interactions. In the context of urban planning, Big Data encompasses information from various sources such as social media, mobile phones, public records, and sensor networks.

2. Internet of Things (IoT):

- IoT refers to the network of physical objects—devices, vehicles, buildings, and other items embedded with sensors, software, and network connectivity—that enable these objects to collect and exchange data. IoT devices in urban environments include smart meters, environmental sensors, connected vehicles, and intelligent streetlights.

Applications of Big Data and IoT in Urban Planning

Applications of Big Data and IoT in urban planning enhance city management through improved data collection, analysis, and real-time decision-making.

Smart Transportation Systems

IoT devices such as GPS-equipped buses, traffic cameras, and connected traffic lights collect real-time data on traffic flow, vehicle speeds, and congestion. This data is analyzed to optimize traffic light timings, manage public transportation schedules, and reduce traffic jams. Predictive analytics can forecast traffic patterns, helping cities to plan for future infrastructure needs.

Energy Management

Smart grids equipped with IoT sensors monitor and manage the distribution of electricity in real time, balancing supply and demand more efficiently. Data from smart meters in homes and businesses can help identify energy consumption patterns, leading to better energy conservation strategies and the integration of renewable energy sources.

Environmental Monitoring

IoT sensors deployed throughout a city can monitor air quality, noise levels, and water quality in real time. This data helps urban planners identify pollution hotspots, implement mitigation strategies, and ensure compliance with environmental regulations. Big Data analytics can also predict environmental trends and assess the impact of urban development on natural resources.

Public Safety and Security

IoT devices such as surveillance cameras, gunshot detectors, and emergency response sensors enhance public safety by providing real-time data to law enforcement and emergency services. Analyzing this data helps identify crime patterns, improve response times, and allocate resources more effectively.

Smart Buildings and Infrastructure

Buildings equipped with IoT sensors can monitor occupancy, energy use, and indoor air quality, optimizing heating, cooling, and lighting systems to enhance comfort and reduce energy consumption. Data from smart infrastructure, such as bridges and roads, can detect wear and tear, enabling proactive maintenance and preventing structural failures.

Urban Planning and Development

Big Data analytics can inform urban planning decisions by providing insights into population growth, migration patterns, and land use trends. This data helps planners design cities that are more sustainable, equitable, and resilient, ensuring that infrastructure and services meet the needs of current and future residents.

Benefits of Big Data and IoT in Urban Planning

1. Enhanced Decision-Making:

- Access to real-time data and advanced analytics enables urban planners to make more informed decisions, resulting in more effective and efficient urban management. Data-driven insights help prioritize investments, optimize resource allocation, and improve the overall quality of urban life.

2. Increased Efficiency:

- IoT-enabled smart systems streamline city operations, reducing waste and lowering operational costs. For example, smart waste management systems can optimize collection routes based on real-time data, reducing fuel consumption and emissions.

3. Improved Sustainability:

- Big Data and IoT technologies support sustainable urban development by optimizing energy use, reducing pollution, and enhancing resource management. Smart grids, renewable energy integration, and efficient public transportation systems contribute to a lower carbon footprint.

4. Greater Resilience:

- Real-time monitoring and predictive analytics enhance a city's ability to respond to emergencies and adapt to changing conditions. Early warning systems for natural disasters, infrastructure health monitoring, and adaptive traffic management increase urban resilience.

5. Enhanced Quality of Life:

- Smart city technologies improve the quality of life for residents by providing better public services, reducing pollution, and enhancing safety. Smart transportation systems, efficient energy management, and clean environments contribute to healthier, more livable urban spaces.

Challenges and Considerations

1. Data Privacy and Security:

- The collection and analysis of vast amounts of data raise concerns about privacy and security. Ensuring that data is collected, stored, and used responsibly, with appropriate safeguards, is essential to maintain public trust and protect sensitive information.

2. Interoperability:

- Integrating diverse IoT devices and data sources can be challenging due to differing standards and protocols. Achieving interoperability requires the development of common frameworks and standards to ensure seamless communication and data exchange.

3. Infrastructure and Investment:

- Implementing IoT and Big Data technologies requires significant investment in infrastructure, including sensors, communication networks, and data centers. Securing funding and demonstrating the long-term value of these investments can be challenging for city governments.

4. Data Management:

- Managing and analyzing large volumes of data requires advanced tools and expertise. Cities need to invest in data management systems and train personnel to leverage data effectively for urban planning.

5. Ethical Considerations:

- The use of data and technology in urban planning must consider ethical implications, such as potential biases in data and algorithms, and the equitable distribution of the benefits of smart city technologies.

Predictive Analytics for Climate Adaptation

Predictive analytics, powered by advanced data analysis techniques and machine learning algorithms, is becoming an indispensable tool in the fight against climate change. By leveraging vast amounts of data, predictive analytics can forecast climate-related events, assess risks, and inform proactive measures to adapt to changing environmental conditions. This section explores the role of predictive analytics in climate adaptation, its applications, benefits, and challenges, and how it can help urban areas prepare for and mitigate the impacts of climate change.

Understanding Predictive Analytics

Predictive analytics involves using historical data, statistical algorithms, and machine learning techniques to identify patterns and make predictions about future events. In the context of climate adaptation, predictive analytics can process and analyze data from various sources, such as weather stations, satellite imagery, and climate models, to forecast climate trends and extreme weather events.

Applications of Predictive Analytics in Climate Adaptation

1. Extreme Weather Forecasting:

- Predictive analytics can enhance the accuracy of weather forecasts by analyzing large datasets and identifying patterns that precede extreme weather events, such as hurricanes, floods, and heatwaves. Early warning systems based on these predictions enable cities to prepare and respond more effectively, minimizing damage and loss of life.

- For example, advanced models can predict the likelihood of hurricanes making landfall in specific regions, allowing for timely evacuations and resource allocation.

2. Flood Risk Management:

- By analyzing rainfall patterns, river levels, and land use data, predictive analytics can identify areas at high risk of flooding. This information helps urban planners design and implement flood mitigation measures, such as improved drainage systems, flood barriers, and green infrastructure.

- Real-time monitoring combined with predictive analytics can also trigger early warnings for communities, enabling them to take preventive actions and reduce flood impacts.

3. Heatwave Prediction and Management:

- Predictive models can forecast heatwaves and their potential severity by analyzing temperature trends, humidity levels, and atmospheric conditions. This information is crucial for cities to implement heat action plans, such as opening cooling centers, issuing public health advisories, and adjusting work schedules.

- Additionally, urban planners can use this data to design heat-resilient infrastructure, such as green roofs, urban forests, and reflective building materials.

4. Agricultural Planning:

- Predictive analytics helps farmers and agricultural planners anticipate changes in climate conditions, such as droughts, frost, and shifts in growing seasons. By forecasting these changes, farmers can adjust their planting schedules, select appropriate crop varieties, and optimize irrigation practices.

- This proactive approach increases agricultural resilience, ensuring food security and reducing economic losses due to climate variability.

5. Water Resource Management:

- Predictive analytics can forecast changes in water availability by analyzing precipitation patterns, snowpack levels, and river flows. This information helps water managers allocate resources efficiently, plan for drought conditions, and implement conservation measures.

- Cities can use predictive models to optimize the operation of reservoirs and water distribution systems, ensuring a stable water supply even during periods of scarcity.

6. Public Health:

- Predictive analytics can identify the potential health impacts of climate change by analyzing data on temperature, humidity, air quality, and disease prevalence. This information helps public health officials prepare for climate-related health issues, such as heat-related illnesses, vector-borne diseases, and respiratory conditions.

- By anticipating health risks, cities can develop targeted interventions, allocate medical resources, and educate communities on preventive measures.

Benefits of Predictive Analytics for Climate Adaptation

1. Proactive Decision-Making:

- Predictive analytics enables cities to anticipate climate impacts and take proactive measures to mitigate risks. This approach reduces the reliance on reactive responses, which are often less effective and more costly.

2. Resource Optimization:

- By providing accurate forecasts, predictive analytics helps optimize the allocation of resources, such as emergency services, medical supplies, and infrastructure investments. This ensures that resources are used efficiently and effectively to address climate risks.

3. Increased Resilience:

- Cities that leverage predictive analytics for climate adaptation are better equipped to withstand and recover from climate-related events. This increased resilience enhances the overall sustainability and livability of urban areas.

4. Cost Savings:

- Proactive climate adaptation measures informed by predictive analytics can prevent or minimize damage from extreme weather events, reducing recovery and reconstruction costs. This results in significant long-term savings for cities and communities.

5. Public Awareness and Engagement:

- Predictive analytics can inform public awareness campaigns and community engagement efforts, educating residents about climate risks and encouraging them to take preventive actions. This fosters a culture of preparedness and resilience within communities.

Challenges of Predictive Analytics for Climate Adaptation

1. Data Quality and Availability:

- The accuracy of predictive models depends on the quality and availability of data. Inconsistent or incomplete data can lead to unreliable predictions, limiting the effectiveness of climate adaptation measures.

2. Technical Expertise:

- Implementing predictive analytics requires specialized knowledge in data science, machine learning, and climate science. Cities may need to invest in training or hire experts to develop and manage predictive models.

3. Infrastructure and Technology:

- Advanced infrastructure and technology are necessary to collect, store, and analyze large datasets. Developing and maintaining this infrastructure can be costly and challenging, particularly for resource-constrained cities.

4. Uncertainty and Complexity:

- Climate systems are inherently complex and influenced by numerous factors, leading to uncertainties in predictions. It is essential to communicate these uncertainties transparently and consider them in decision-making processes.

5. Interdisciplinary Collaboration:

- Effective climate adaptation requires collaboration across various disciplines, including urban planning, public health, environmental science, and emergency management. Coordinating efforts and integrating diverse perspectives can be challenging but is crucial for successful outcomes.

Citizen Engagement and Participatory Planning

Citizen engagement and participatory planning are crucial components of sustainable urban development. Involving residents in the planning process ensures that urban development meets the needs and aspirations of the community, fosters a sense of ownership, and enhances the legitimacy and effectiveness of urban policies. This section explores the importance of citizen engagement, the methods and tools used to facilitate participatory planning, and the benefits and challenges associated with these practices.

Importance of Citizen Engagement

1. Inclusivity and Equity:

- Engaging citizens in urban planning helps ensure that the voices of diverse community members are heard, including marginalized and vulnerable groups. This inclusivity leads to more equitable outcomes and addresses the needs of all residents.

2. Enhanced Decision-Making:

- Citizens provide valuable local knowledge and insights that can improve the quality and relevance of planning decisions. Their involvement can help identify potential issues and opportunities that planners might overlook.

3. Community Ownership and Trust:

- When residents actively participate in planning processes, they are more likely to support and advocate for the outcomes. This sense of ownership fosters community pride and trust in local government.

4. Transparency and Accountability:

- Participatory planning promotes transparency by making decision-making processes more open and accessible. It also holds planners and policymakers accountable to the community they serve.

Methods and Tools for Participatory Planning

1. Public Meetings and Workshops:

- Public meetings and workshops provide forums for residents to share their ideas, concerns, and feedback on urban planning initiatives. These events can be structured to include presentations, Q&A sessions, and group discussions, fostering active participation and dialogue.

2. Surveys and Questionnaires:

- Surveys and questionnaires are effective tools for gathering input from a broad segment of the community. They can be distributed online or in person, allowing residents to provide feedback on specific issues or proposed projects.

3. Focus Groups:

- Focus groups involve small, diverse groups of residents who discuss specific topics in depth. These sessions can uncover detailed insights and foster a deeper understanding of community perspectives.

4. Participatory Mapping:

- Participatory mapping involves residents in creating maps that highlight community assets, challenges, and priorities. This visual tool helps planners understand spatial aspects of the community and identify areas for intervention.

5. Online Platforms and Social Media:

- Online platforms and social media offer accessible and convenient ways for residents to engage in the planning process. Tools such as interactive maps, discussion forums, and virtual town halls can broaden participation and reach a wider audience.

6. Citizen Advisory Committees:

- Citizen advisory committees consist of community representatives who provide ongoing input and guidance on planning initiatives. These committees ensure continuous engagement and facilitate communication between residents and planners.

Benefits of Citizen Engagement and Participatory Planning

1. Improved Planning Outcomes:

- By incorporating diverse perspectives and local knowledge, participatory planning leads to more informed and contextually relevant decisions. This results in better-designed projects that are more likely to succeed and meet community needs.

2. Strengthened Community Relationships:

- Engaging residents in planning processes builds stronger relationships between community members and local government. It fosters collaboration, mutual understanding, and a shared vision for the future.

3. Increased Civic Participation:

- Participatory planning encourages residents to become more involved in civic life and local governance. This heightened civic engagement can lead to greater community resilience and collective action.

4. Enhanced Social Capital:

- Through participatory processes, residents develop networks, skills, and a sense of agency. These social capital assets contribute to the overall vitality and cohesion of the community.

Challenges of Citizen Engagement and Participatory Planning

1. Resource Intensive:

- Effective citizen engagement requires significant time, effort, and resources. Organizing meetings, conducting surveys, and managing online platforms can be resource-intensive, particularly for smaller municipalities with limited budgets.

2. Ensuring Inclusive Participation:

- Engaging a diverse cross-section of the community can be challenging. Planners must proactively reach out to underrepresented groups and address barriers to participation, such as language, accessibility, and time constraints.

3. Managing Conflicting Interests:

- Communities are diverse, and different groups may have conflicting interests and priorities. Facilitating constructive dialogue and finding common ground requires skilled mediation and negotiation.

4. Maintaining Momentum:

- Sustaining citizen engagement over the long term can be difficult. Continuous communication, feedback loops, and demonstrating the impact of citizen input are essential to keep residents motivated and involved.

5. Balancing Expertise and Lay Input:

- Integrating professional expertise with community input can be complex. Planners must balance technical considerations with residents' perspectives, ensuring that decisions are both technically sound and socially acceptable.

Chapter 8: Climate Adaptation Strategies for Cities

As the impacts of climate change become increasingly evident, cities around the world are facing unprecedented challenges. From rising sea levels and extreme weather events to heatwaves and resource scarcity, urban areas must develop robust strategies to adapt to these changing conditions. This chapter explores various climate adaptation strategies that cities can implement to enhance their resilience and sustainability. We will examine approaches such as green infrastructure, resilient building designs, water management systems, and community engagement initiatives. By integrating these strategies, cities can better protect their residents, infrastructure, and ecosystems from the adverse effects of climate change, ensuring a sustainable and resilient urban future.

Urban Heat Island Mitigation

Urban Heat Islands (UHIs) are a common phenomenon in densely populated areas where the concentration of buildings, roads, and other infrastructure leads to higher temperatures compared to surrounding rural areas. This effect is caused by the absorption and retention of heat by materials such as asphalt, concrete, and rooftops, which release heat slowly, elevating nighttime temperatures. UHIs exacerbate heatwaves, increase energy consumption, and pose health risks to urban residents. Effective UHI mitigation strategies are essential for creating more livable, sustainable, and resilient cities. This section explores the causes of UHIs, their impacts, and various mitigation strategies that can be implemented.

Causes of Urban Heat Islands

1. Building Materials:

- Urban areas are often constructed with materials that have high thermal mass, such as concrete and asphalt. These materials absorb and retain heat during the day and release it slowly at night, contributing to elevated temperatures.

2. Lack of Vegetation:

- Vegetation cools the air through the process of evapotranspiration. Urban areas with sparse green spaces and tree cover lack this natural cooling effect, leading to higher temperatures.

3. Human Activities:

- Activities such as transportation, industrial processes, and energy consumption generate heat, further contributing to the UHI effect. Air conditioners, while cooling indoor environments, expel heat into the surrounding atmosphere.

4. Urban Geometry:

- The layout and density of buildings can influence air circulation. Narrow streets and tall buildings can trap heat and reduce airflow, preventing the dispersal of warm air.

Impacts of Urban Heat Islands

1. Increased Energy Consumption:

- Higher temperatures in urban areas lead to increased use of air conditioning, which in turn raises energy demand and utility costs. This increased energy consumption can strain power grids, especially during peak periods.

2. Health Risks:

- Prolonged exposure to high temperatures can lead to heat-related illnesses such as heatstroke, dehydration, and respiratory problems.

Vulnerable populations, including the elderly, children, and those with pre-existing health conditions, are at greater risk.

3. Decreased Air Quality:

- Elevated temperatures can accelerate the formation of ground-level ozone, a harmful air pollutant. Poor air quality can exacerbate respiratory conditions and reduce overall public health.

4. Water Quality Issues:

- Higher temperatures can increase the temperature of water bodies, affecting aquatic ecosystems. Warm waters can lead to the growth of harmful algae blooms and reduce oxygen levels, impacting fish and other aquatic life.

Mitigation Strategies for Urban Heat Islands

1. Increasing Urban Green Spaces:

- Parks and Green Roofs: Expanding parks, green roofs, and green walls can significantly reduce UHI effects. These green spaces provide shade, reduce surface temperatures, and enhance evapotranspiration, cooling the surrounding air.

- Tree Planting: Planting trees along streets, in parks, and on private properties can provide shade and lower ambient temperatures. Trees also improve air quality and contribute to biodiversity.

2. Cool Roofs and Pavements:

- Cool Roofs: Installing cool roofs that reflect more sunlight and absorb less heat than standard roofs can reduce building temperatures and energy use. These roofs are coated with reflective materials or painted in light colors to enhance their cooling effect.

- Cool Pavements: Using materials with higher reflectivity for pavements and roads can help lower surface temperatures. Cool pavements can be made from materials such as concrete, reflective coatings, and permeable surfaces.

3. Urban Design and Planning:

- Building Orientation and Design: Designing buildings to maximize natural ventilation and minimize heat absorption can help reduce indoor and outdoor temperatures. Features such as shading devices, green facades, and reflective surfaces can enhance cooling.

- Urban Layout: Planning urban layouts to enhance airflow and reduce heat accumulation is crucial. Wide streets, open spaces, and building setbacks can improve ventilation and mitigate UHI effects.

4. Water Features:

- Urban Water Bodies: Integrating water features such as fountains, ponds, and artificial lakes into urban areas can provide cooling through evaporation. These features also enhance the aesthetic appeal and recreational value of cities.

- Rain Gardens and Bioswales: Implementing rain gardens and bioswales can help manage stormwater while providing cooling benefits. These green infrastructure elements absorb and filter runoff, reducing surface temperatures.

5. Community Engagement and Education:

- Public Awareness Campaigns: Educating residents about the causes and impacts of UHIs and encouraging participation in mitigation efforts is essential. Community-driven initiatives such as tree planting drives and green roof installations can amplify impact.

- Incentive Programs: Offering incentives for adopting UHI mitigation measures, such as tax rebates for installing cool roofs or planting trees, can motivate residents and businesses to contribute to cooling efforts.

Benefits of Urban Heat Island Mitigation

1. Reduced Energy Costs:

- Implementing UHI mitigation strategies can lower energy consumption, reducing utility bills for residents and businesses. This also alleviates pressure on power grids, particularly during peak demand periods.

2. Improved Public Health:

- Lowering urban temperatures can reduce the incidence of heat-related illnesses and improve overall public health. Enhanced air quality resulting from reduced temperatures and increased vegetation also benefits respiratory health.

3. Enhanced Livability:

- Cooler urban environments improve the quality of life for residents by providing comfortable outdoor spaces, reducing stress, and promoting physical activity. Green spaces and water features also enhance the aesthetic appeal of cities.

4. Environmental Benefits:

- UHI mitigation contributes to broader environmental goals by reducing greenhouse gas emissions, conserving water, and promoting biodiversity. Green spaces and cool roofs can sequester carbon and support urban wildlife.

Flood Management and Coastal Protection

As climate change continues to intensify, the frequency and severity of flooding and coastal erosion have increased, posing significant risks to urban areas. Effective flood management and coastal protection strategies are essential for safeguarding communities, infrastructure, and ecosystems. This section explores the causes and impacts of flooding and coastal erosion, and outlines various strategies and technologies for mitigating these risks.

Causes and Impacts of Flooding

Flooding can result from various factors, including heavy rainfall, storm surges, river overflow, and inadequate drainage systems. Urban areas are particularly vulnerable due to high levels of impervious surfaces, such as roads and buildings, which prevent water absorption and exacerbate runoff.

Impacts of Flooding:

- Infrastructure Damage: Floodwaters can cause severe damage to buildings, roads, bridges, and utility systems, leading to costly repairs and disruptions.

- Public Health Risks: Flooding can contaminate water supplies, spread waterborne diseases, and create breeding grounds for pests, posing serious health risks to residents.

- Economic Losses: Businesses may face significant financial losses due to property damage, inventory destruction, and interruptions in operations. Flooding also affects agriculture, leading to crop losses and food shortages.

- Environmental Degradation: Floods can erode soil, destroy habitats, and wash pollutants into waterways, harming aquatic ecosystems and biodiversity.

Causes and Impacts of Coastal Erosion

Coastal erosion is primarily driven by natural processes such as wave action, tides, and currents, but human activities like construction, sand mining, and deforestation can accelerate it. Climate change exacerbates coastal erosion through rising sea levels and increased storm intensity.

Impacts of Coastal Erosion:

- Loss of Land: Erosion can result in the loss of valuable land, affecting properties, infrastructure, and natural habitats along the coast.

- Infrastructure Damage: Coastal erosion can undermine the foundations of buildings, roads, and seawalls, leading to structural failures and costly repairs.

- Displacement of Communities: Erosion can force coastal communities to relocate, disrupting lives and livelihoods.

- Ecological Damage: Coastal ecosystems, such as mangroves, coral reefs, and wetlands, can be severely impacted by erosion, leading to loss of biodiversity and ecosystem services.

Flood Management Strategies

1. Green Infrastructure:

- Rain Gardens and Bioswales: These landscaped areas are designed to capture, filter, and infiltrate stormwater, reducing runoff and mitigating flood risks. Rain gardens and bioswales enhance groundwater recharge and improve water quality.

- Green Roofs and Walls: Vegetated roofs and walls absorb rainwater, reduce runoff, and provide insulation, contributing to flood management and energy efficiency.

2. Improved Drainage Systems:

- Permeable Pavements: These surfaces allow water to infiltrate the ground, reducing runoff and easing pressure on drainage systems. Permeable pavements can be used in parking lots, sidewalks, and roads.

- Upgraded Stormwater Systems: Modernizing and expanding stormwater infrastructure, such as culverts, drains, and retention basins, can enhance the capacity to manage heavy rainfall and prevent flooding.

3. Flood Barriers and Levees:

- Temporary Barriers: Deployable barriers, such as sandbags and water-filled dams, can provide temporary protection against floodwaters during emergencies.

- Permanent Levees and Dikes: Constructing permanent levees and dikes along rivers and coastlines can offer long-term flood protection. These structures must be properly maintained and monitored to ensure their effectiveness.

4. Floodplain Management:

- Zoning and Land Use Planning: Implementing zoning regulations and land use policies that restrict development in flood-prone areas can reduce vulnerability and minimize damage. Establishing floodplain restoration projects can enhance natural flood storage and habitat conservation.

- Relocation and Buyout Programs: Offering incentives for relocating properties out of high-risk flood zones can reduce exposure and long-term costs associated with flood damage.

Coastal Protection Strategies

1. Natural Barriers:

- Mangroves and Wetlands: Restoring and protecting mangroves and wetlands can provide natural buffers against storm surges and coastal erosion. These ecosystems also support biodiversity and enhance carbon sequestration.

- Coral Reefs and Oyster Beds: Healthy coral reefs and oyster beds can dissipate wave energy, reducing the impact of storm surges and erosion. Conservation and restoration efforts can enhance these natural defenses.

2. Engineered Solutions:

- Seawalls and Revetments: Constructing seawalls and revetments along vulnerable coastlines can provide immediate protection against erosion and storm surges. These structures should be designed to withstand harsh marine conditions and minimize environmental impact.

- Breakwaters and Groynes: Installing breakwaters and groynes can interrupt wave energy and stabilize beaches, reducing erosion and protecting coastal properties.

3. Beach Nourishment:

- Sand Dredging and Deposition: Beach nourishment involves dredging sand from offshore sources and depositing it on eroding beaches. This strategy can provide temporary relief from erosion and enhance recreational opportunities.

- Dune Restoration: Restoring and stabilizing sand dunes with vegetation can protect inland areas from storm surges and erosion. Dune restoration projects often involve planting native grasses and installing sand fences.

4. Integrated Coastal Zone Management (ICZM):

- Holistic Planning: ICZM involves coordinating the management of coastal areas across sectors and stakeholders, integrating environmental, social, and economic objectives. This approach ensures sustainable development and long-term resilience.

- Community Engagement: Engaging local communities in coastal protection efforts can enhance awareness, foster stewardship, and ensure the success of management strategies.

Benefits of Flood Management and Coastal Protection

1. Reduced Damage and Costs:

- Implementing effective flood management and coastal protection measures can significantly reduce the damage to infrastructure, properties, and ecosystems, leading to lower repair and recovery costs.

2. Enhanced Public Safety:

- By mitigating flood and erosion risks, these strategies protect communities from the dangers of extreme weather events, reducing injuries, fatalities, and health impacts.

3. Environmental Preservation:

- Integrating natural solutions into flood and coastal management enhances biodiversity, improves water quality, and supports ecosystem services such as carbon sequestration and habitat provision.

4. Increased Resilience:

- Building resilience to flooding and coastal erosion ensures that urban areas can adapt to and recover from the impacts of climate change, maintaining their functionality and livability.

Enhancing Urban Resilience through Design

Urban resilience is the ability of cities to withstand, adapt to, and recover from various shocks and stresses, including those induced by climate change, natural disasters, and socio-economic challenges. Enhancing urban resilience through design involves integrating sustainable and adaptive strategies into the urban fabric to create robust, flexible, and enduring cities. This section explores the principles and practices that underpin resilient urban design, focusing on various aspects such as infrastructure, building design, public spaces, and community engagement.

Principles of Resilient Urban Design

Resilient urban design is grounded in principles that prioritize flexibility, sustainability, and inclusivity. These principles guide the development of urban spaces that can adapt to changing conditions and withstand potential disruptions.

Flexibility and Adaptability:

- Urban design should incorporate flexible and adaptive features that allow spaces and structures to be easily modified in response to evolving needs and conditions. This includes modular construction, multifunctional spaces, and adaptable infrastructure systems.

Sustainability:

- Resilient design integrates sustainable practices that reduce environmental impact and enhance resource efficiency. This involves using eco-friendly materials, incorporating green infrastructure, and promoting energy efficiency.

Inclusivity:

- Ensuring that urban design is inclusive and accessible to all members of the community is vital for resilience. This means considering the needs of diverse populations, including vulnerable groups, and fostering social equity.

Resilient Infrastructure

Infrastructure systems are the backbone of urban resilience, providing essential services and connectivity. Designing resilient infrastructure involves ensuring that these systems can withstand and quickly recover from disruptions.

Robust and Redundant Systems:

- Infrastructure should be designed with robustness and redundancy to ensure continuous operation during adverse conditions. This includes building redundancy into power grids, water supply networks, and transportation systems to provide backup options in case of failures.

Smart Technologies:

- Integrating smart technologies into infrastructure systems enhances their resilience by enabling real-time monitoring, early warning, and rapid response capabilities. Smart grids, sensor networks, and automated controls can optimize the performance and adaptability of urban infrastructure.

Decentralized Systems:

- Decentralizing infrastructure systems, such as energy production and water management, can enhance resilience by reducing dependency on centralized sources and enabling localized solutions. Decentralized systems are often more adaptable and can recover more quickly from disruptions.

Resilient Building Design

Buildings play a crucial role in urban resilience, providing shelter and safety for residents. Resilient building design focuses on creating structures that can withstand environmental stresses and adapt to changing conditions.

Durable Materials and Construction:

- Using durable materials and construction techniques that enhance the structural integrity of buildings is essential for resilience. This includes designing buildings to withstand extreme weather events, such as hurricanes, earthquakes, and floods.

Energy Efficiency and Renewable Energy:

- Incorporating energy-efficient features and renewable energy sources into building designs reduces dependency on external power supplies and enhances resilience to energy disruptions. Features such as solar panels, efficient insulation, and passive cooling systems contribute to energy resilience.

Flexible and Multifunctional Spaces:

- Designing buildings with flexible and multifunctional spaces allows them to be easily reconfigured for different uses and conditions. This adaptability is crucial for responding to changing needs, such as converting spaces for emergency shelter or community support during crises.

Resilient Public Spaces

Public spaces are vital for fostering community resilience and well-being. Designing resilient public spaces involves creating areas that support social interaction, provide safe havens, and contribute to environmental sustainability.

Green Infrastructure:

- Integrating green infrastructure, such as parks, green roofs, and urban forests, into public spaces enhances resilience by providing natural cooling, stormwater management, and biodiversity benefits. These spaces also offer recreational and mental health benefits for residents.

Safe and Inclusive Design:

- Public spaces should be designed to be safe, accessible, and inclusive for all community members. This includes providing adequate lighting, clear signage, and barrier-free access to ensure that everyone can use and enjoy these spaces.

Community Gathering Spaces:

- Creating spaces that facilitate community gatherings and social interaction strengthens social cohesion and resilience. Public plazas, community gardens, and cultural centers provide opportunities for residents to connect, share resources, and support each other during crises.

Community Engagement and Participation

Engaging the community in the design and development of urban spaces is essential for resilience. Participatory planning processes ensure that the needs and preferences of residents are considered, fostering a sense of ownership and collective responsibility.

Inclusive Planning Processes:

- Involving diverse community members in planning and decision-making processes ensures that the voices of all groups are heard and respected. This inclusivity helps identify vulnerabilities and prioritize resilience measures that benefit the entire community.

Education and Awareness:

- Educating residents about resilience and sustainable practices empowers them to contribute to urban resilience efforts. Public awareness campaigns, workshops, and educational programs can enhance community understanding and involvement.

Local Partnerships and Collaboration:

- Building partnerships between local government, businesses, non-profits, and community organizations enhances the capacity for resilience. Collaborative efforts leverage diverse resources and expertise to implement effective resilience strategies.

Policies and Frameworks for Climate Adaptation

As climate change continues to pose significant threats to urban environments, the development and implementation of effective policies and frameworks for climate adaptation are critical. These strategies provide the foundation for cities to prepare for, respond to, and recover from climate-related impacts. This section explores various policies and frameworks that support climate adaptation, including international agreements, national policies, and local strategies.

International Agreements and Initiatives

International agreements and initiatives play a crucial role in setting global standards and fostering collaboration among countries to address climate change.

The Paris Agreement:

- The Paris Agreement is a landmark international treaty adopted in 2015 under the United Nations Framework Convention on Climate Change (UNFCCC). It aims to limit global warming to well below 2 degrees Celsius, with efforts to limit it to 1.5 degrees Celsius. The agreement encourages countries to develop and implement national adaptation plans and enhance their resilience to climate impacts.

The Sendai Framework for Disaster Risk Reduction:

- Adopted in 2015, the Sendai Framework for Disaster Risk Reduction outlines strategies to reduce disaster risk and enhance resilience. It emphasizes the importance of understanding disaster risk, strengthening disaster risk governance, and investing in disaster risk reduction for resilience. The framework encourages countries to integrate climate adaptation and disaster risk reduction into their policies and planning.

National Policies and Strategies

National governments play a key role in establishing policies and strategies that guide climate adaptation efforts at all levels of government.

National Adaptation Plans (NAPs):

- National Adaptation Plans (NAPs) are strategic frameworks developed by countries to identify and address their specific climate adaptation needs and priorities. NAPs provide a roadmap for integrating climate adaptation into national development planning and policy-making, ensuring a coordinated and comprehensive approach to building resilience.

Climate Action Plans:

- Many countries have developed Climate Action Plans that outline specific measures to reduce greenhouse gas emissions and adapt to climate change. These plans often include sector-specific strategies, such as enhancing water resource management, improving agricultural resilience, and strengthening coastal protection. Climate Action Plans provide a clear framework for implementing adaptation measures and monitoring progress.

Local Policies and Regulations

Local governments are at the forefront of climate adaptation efforts, as they have direct influence over urban planning, infrastructure development, and community engagement.

Municipal Climate Adaptation Plans:

- Municipal Climate Adaptation Plans are developed by local governments to address the unique climate risks and vulnerabilities of their communities. These plans outline specific actions to enhance resilience, such as upgrading drainage systems, implementing green infrastructure, and promoting sustainable building practices. Local adaptation plans ensure that climate adaptation is integrated into urban development and policy-making.

Zoning and Land Use Regulations:

- Zoning and land use regulations play a critical role in reducing climate risks and enhancing resilience. Local governments can implement zoning codes that restrict development in flood-prone areas, encourage the preservation of green spaces, and promote the use of climate-resilient building materials. These regulations help guide sustainable urban growth and minimize the impacts of climate change on communities.

Frameworks for Collaborative Action

Effective climate adaptation requires collaboration across multiple levels of government, sectors, and stakeholders. Frameworks for collaborative action facilitate coordinated efforts and shared responsibility.

Public-Private Partnerships (PPPs):

- Public-Private Partnerships (PPPs) bring together government agencies, private sector entities, and non-profit organizations to implement climate adaptation projects. PPPs leverage the resources, expertise, and innovation of diverse partners to enhance the effectiveness and scalability of adaptation efforts. Examples of PPP initiatives include the development of resilient infrastructure, green building programs, and community-based adaptation projects.

Community-Based Adaptation (CBA):

- Community-Based Adaptation (CBA) involves engaging local communities in the planning and implementation of climate adaptation strategies. CBA recognizes the valuable knowledge and resources that communities possess and empowers them to take an active role in building resilience. This approach ensures that adaptation measures are locally relevant, culturally appropriate, and sustainable.

Monitoring and Evaluation Frameworks

Monitoring and evaluation frameworks are essential for assessing the effectiveness of climate adaptation policies and identifying areas for improvement.

Indicators and Metrics:

- Developing indicators and metrics to measure the progress and outcomes of climate adaptation efforts is crucial for monitoring success and making data-driven decisions. These indicators can include metrics related to infrastructure resilience, ecosystem health, and community preparedness. Regular monitoring and reporting help ensure transparency and accountability in adaptation initiatives.

Adaptive Management:

- Adaptive management is an iterative process that involves regularly assessing the effectiveness of adaptation measures and making adjustments as needed. This approach allows for flexibility and learning, enabling governments and stakeholders to respond to new information and changing conditions. Adaptive management ensures that climate adaptation strategies remain effective and relevant over time.

Chapter 9: Financing and Policy Frameworks

Securing adequate financing and establishing robust policy frameworks are critical components of effective climate adaptation and mitigation efforts. As cities and nations grapple with the escalating impacts of climate change, the need for innovative funding mechanisms and supportive policies becomes ever more pressing. This chapter delves into the various financial instruments and policy strategies that can drive sustainable urban development and enhance climate resilience. We will explore international, national, and local perspectives, highlighting successful case studies and best practices. By understanding the interplay between financing and policy, stakeholders can better navigate the complexities of funding climate initiatives and implementing effective adaptation measures.

Funding Mechanisms for Urban Innovation

Securing financial resources is essential for implementing innovative solutions that enhance urban resilience and sustainability. Various funding mechanisms are available to support urban innovation, each offering unique benefits and challenges. This section explores different funding sources and strategies, including government grants, private investments, public-private partnerships, and innovative financing tools.

Government Grants and Subsidies

Government grants and subsidies are critical for kickstarting urban innovation projects, particularly those that may not attract immediate private investment due to high risk or long-term returns.

National and Local Government Grants:

- National and local governments often provide grants and subsidies to support urban innovation initiatives. These funds can be allocated to projects such as renewable energy installations, smart grid development, and sustainable transportation systems. Government grants help reduce the financial burden on municipalities and encourage the adoption of cutting-edge technologies.

Research and Development (R&D) Funding:

- Governments also allocate funds specifically for research and development to foster innovation in urban planning and infrastructure. R&D funding supports pilot projects, feasibility studies, and the development of new technologies, paving the way for broader implementation of innovative solutions.

Private Investments

Private investments play a vital role in financing urban innovation, bringing in capital, expertise, and market-driven efficiencies.

Venture Capital and Private Equity:

- Venture capital (VC) and private equity (PE) firms invest in early-stage companies and projects with high growth potential. Urban innovation projects, such as smart city technologies, renewable energy startups, and sustainable real estate developments, often attract VC and PE funding. These investors provide not only financial resources but also strategic guidance to help scale innovative solutions.

Corporate Investments:

- Corporations are increasingly investing in urban innovation as part of their sustainability and corporate social responsibility (CSR) initiatives. Companies may fund projects that align with their business interests, such as energy-efficient building technologies, electric vehicle infrastructure, and smart logistics solutions. Corporate investments can drive significant advancements in urban innovation.

Public-Private Partnerships (PPPs)

PPPs combine the strengths of the public and private sectors to finance and implement urban innovation projects.

Infrastructure Development:

- PPPs are commonly used to finance large-scale infrastructure projects, such as transportation networks, water treatment facilities, and energy systems. These partnerships leverage private sector efficiency and innovation while ensuring public sector oversight and accountability. Successful PPPs can accelerate the deployment of advanced infrastructure and technologies.

Community Development:

- PPPs can also support community development projects that enhance urban resilience and sustainability. Examples include affordable housing initiatives, green space development, and community energy projects. By aligning public and private interests, PPPs can address social and environmental challenges in urban areas.

Innovative Financing Tools

Innovative financing tools offer alternative methods for funding urban innovation, often involving creative approaches to capital formation and risk management.

Green Bonds:

- Green bonds are debt instruments specifically earmarked for environmentally friendly projects. Cities and municipalities can issue green bonds to raise funds for initiatives such as renewable energy installations, energy-efficient buildings, and sustainable transportation systems. Green bonds attract socially responsible investors and provide a dedicated funding stream for sustainability projects.

Impact Investing:

- Impact investing involves investing in projects or companies that generate measurable social and environmental benefits alongside financial returns. Impact investors seek to support initiatives that address urban challenges, such as affordable housing, clean energy, and sustainable agriculture. By prioritizing positive impacts, impact investing mobilizes capital for transformative urban projects.

Crowdfunding:

- Crowdfunding platforms allow individuals and organizations to contribute small amounts of money to support urban innovation projects. This approach can democratize funding and engage the community in project development. Crowdfunding is particularly useful for grassroots initiatives, pilot projects, and community-based solutions.

Climate Finance:

- Climate finance refers to funding specifically aimed at addressing climate change mitigation and adaptation. Sources of climate finance include international funds, such as the Green Climate Fund (GCF) and the Adaptation Fund, as well as bilateral and multilateral aid programs. Climate finance supports projects that enhance urban resilience to climate impacts, such as flood management systems, climate-resilient infrastructure, and disaster preparedness programs.

Challenges and Considerations

While various funding mechanisms are available, securing adequate financing for urban innovation projects can be challenging. Key considerations include:

Project Viability and Risk:

- Investors and funders require assurance that projects are viable and capable of delivering the intended benefits. Conducting thorough feasibility studies, risk assessments, and cost-benefit analyses is essential to attract funding.

Regulatory Environment:

- Navigating regulatory requirements and ensuring compliance with local, national, and international standards can impact project funding and implementation. Understanding the regulatory landscape and engaging with policymakers early in the project development process is crucial.

Stakeholder Engagement:

- Effective stakeholder engagement is vital for securing support and funding for urban innovation projects. This involves building strong partnerships, communicating the benefits of the projects, and addressing stakeholder concerns and interests.

Scalability and Replicability:

- Funders often look for projects that can be scaled and replicated in other urban areas. Demonstrating the scalability and broader applicability of innovative solutions can enhance their attractiveness to investors and funding agencies.

Public-Private Partnerships

PPPs are collaborative arrangements between government entities and private sector companies designed to finance, build, and operate projects that serve the public interest. By combining public oversight with private efficiency and innovation, PPPs offer a viable solution for addressing complex urban challenges and promoting sustainable development. This section explores the various aspects of PPPs, including their structure, benefits, challenges, and best practices for successful implementation.

Structure of Public-Private Partnerships

The structure of a PPP defines the roles, responsibilities, and risk-sharing arrangements between the public and private partners.

Contractual Arrangements

- PPPs are typically formalized through detailed contractual agreements that outline the scope of the project, the financial contributions of each party, the distribution of risks and rewards, and the performance standards to be met. Common types of PPP contracts include build-operate-transfer (BOT), design-build-finance-operate (DBFO), and lease-develop-operate (LDO) agreements.

Risk Allocation

Effective PPPs involve a balanced allocation of risks between the public and private sectors. Generally, the public sector retains risks associated with regulatory changes and public opposition, while the private sector assumes risks related to construction, financing, and operational efficiency. Proper risk allocation is crucial to ensuring that each party is incentivized to perform optimally.

Financing Mechanisms

- Financing for PPP projects can come from a mix of public funds, private investments, and loans. Public funding might include grants, subsidies, and tax incentives, while private financing can involve equity contributions, debt financing, and capital market instruments. The financial structure should ensure project viability and sustainability.

Benefits of Public-Private Partnerships

PPPs offer numerous benefits that can enhance the delivery and management of urban infrastructure and services.

Efficiency and Innovation

- Private sector participation brings efficiency and innovation to project delivery and management. Private companies often have access to advanced technologies, management expertise, and streamlined processes that can improve project outcomes and reduce costs.

Access to Capital

- PPPs enable the mobilization of private capital, which can be crucial for funding large-scale infrastructure projects that might be beyond the financial capacity of public budgets alone. This access to additional resources accelerates project implementation and completion.

Risk Sharing:

- By distributing risks between public and private entities, PPPs reduce the financial and operational burden on any single party. This shared risk framework encourages accountability and performance, as both sectors have a vested interest in the project's success.

Enhanced Public Services

- PPPs can lead to the development of high-quality public services and infrastructure, benefiting communities. Improved transportation networks, water and sanitation systems, healthcare facilities, and educational institutions are examples of public services that can be enhanced through PPPs.

Challenges of Public-Private Partnerships

While PPPs offer significant advantages, they also present challenges that must be managed to ensure their success.

Complexity and Costs

- The complexity of structuring, negotiating, and managing PPPs can lead to high transaction costs. Detailed contracts, extensive legal and financial due diligence, and ongoing oversight require significant resources and expertise.

Regulatory and Political Risks

- Changes in regulatory frameworks or political climates can impact the stability and success of PPP projects. Private investors may be wary of political risks, such as policy shifts, changes in leadership, or public opposition that could affect project continuity and profitability.

Performance Monitoring

- Ensuring that private partners meet agreed-upon performance standards can be challenging. Effective monitoring and enforcement mechanisms must be in place to ensure that projects are delivered on time, within budget, and to the required quality standards.

Equity and Accessibility

- There is a risk that PPP projects may prioritize profitability over equity, potentially leading to disparities in service access. It is essential to incorporate safeguards that ensure all community members benefit equitably from PPP initiatives.

Best Practices for Successful Implementation

Adopting best practices can enhance the effectiveness of PPPs and mitigate associated challenges.

Clear Objectives and Planning

Establishing clear, achievable objectives and thorough planning are essential for successful PPPs. A detailed feasibility study should assess the project's viability, potential risks, and expected benefits, providing a solid foundation for the partnership.

Transparent Procurement Processes

Transparent and competitive procurement processes are vital for attracting qualified private partners and ensuring fair and accountable project delivery. Open bidding and clear evaluation criteria can enhance credibility and stakeholder trust.

Robust Legal and Regulatory Frameworks

Strong legal and regulatory frameworks provide the necessary support and stability for PPP projects. These frameworks should address contract enforcement, dispute resolution, and protection of investor interests, reducing uncertainties and enhancing confidence.

Stakeholder Engagement

Engaging stakeholders, including the community, businesses, and interest groups, ensures that PPP projects align with public needs and priorities. Regular communication and inclusive participation can build support and mitigate potential conflicts.

Performance-Based Contracts

Performance-based contracts that link payments and incentives to the achievement of specific milestones and quality standards can drive better project outcomes. These contracts should include clear metrics, monitoring procedures, and penalties for non-compliance.

Capacity Building

Building the capacity of public sector entities to manage PPPs effectively is crucial. This includes training staff, developing expertise in contract management, and establishing dedicated PPP units to oversee project implementation.

International and National Policy Initiatives

International and national policy initiatives are fundamental in shaping the landscape of climate adaptation and sustainable urban development. These initiatives set the frameworks and guidelines that drive local actions and innovations. This section highlights key international agreements and national policies that support urban resilience and climate adaptation efforts.

International Agreements

International agreements provide a global framework for countries to collaborate on climate adaptation and sustainability goals.

The Paris Agreement

The Paris Agreement, adopted in 2015, is a landmark international accord under the United Nations Framework Convention on Climate Change (UNFCCC). It aims to limit global temperature rise to well below 2 degrees Celsius and encourages countries to enhance their adaptive capacities and resilience to climate impacts through Nationally Determined Contributions (NDCs).

The Sendai Framework for Disaster Risk Reduction

The Sendai Framework for Disaster Risk Reduction, adopted in 2015, focuses on reducing disaster risk and enhancing resilience. It emphasizes the importance of understanding disaster risk, strengthening disaster risk governance, and investing in disaster risk reduction measures to safeguard communities from the impacts of climate change and natural disasters.

National Policies

National policies play a crucial role in translating international commitments into actionable strategies and regulations at the country level.

National Adaptation Plans (NAPs)

National Adaptation Plans (NAPs) are strategic frameworks developed by countries to identify their specific climate adaptation needs and priorities. These plans guide national efforts to integrate climate adaptation into development planning and policy-making, ensuring a coordinated and comprehensive approach to enhancing resilience.

Nationally Determined Contributions (NDCs)

As part of the Paris Agreement, countries submit Nationally Determined Contributions (NDCs), outlining their commitments to reduce greenhouse gas emissions and adapt to climate change. NDCs include specific targets, measures, and policies that reflect each country's unique circumstances and priorities, driving national efforts to mitigate and adapt to climate impacts.

Regional Initiatives

Regional initiatives facilitate collaboration and knowledge-sharing among neighboring countries facing similar climate challenges.

European Union Climate Policies

The European Union (EU) has established comprehensive climate policies, including the European Green Deal, which aims to make Europe climate-neutral by 2050. The EU's adaptation strategy encourages member states to develop national adaptation plans, integrate climate considerations into all policy areas, and promote resilience through regional cooperation.

African Union Climate Strategies

The African Union (AU) implements regional climate strategies, such as the African Union Climate Change Strategy and Action Plan. These initiatives focus on enhancing the adaptive capacity of African countries, promoting sustainable development, and fostering regional cooperation to address shared climate challenges.

Local Implementation

Local governments play a critical role in implementing international and national policies through localized actions and initiatives.

Municipal Climate Action Plans

Municipal Climate Action Plans translate national and international climate goals into specific, actionable measures at the city level. These plans address local climate risks, set adaptation and mitigation targets, and outline strategies for building urban resilience.

Community-Based Adaptation (CBA)

Community-Based Adaptation (CBA) involves engaging local communities in the planning and implementation of climate adaptation strategies. CBA ensures that adaptation measures are locally relevant, culturally appropriate, and sustainable, empowering communities to build their resilience to climate impacts.

Barriers and Opportunities for Implementation

Implementing climate adaptation and sustainability initiatives in urban areas presents a range of challenges and opportunities. Understanding these barriers and opportunities is essential for developing effective strategies and achieving long-term resilience. This section examines the key obstacles to implementation and highlights potential opportunities for overcoming them.

Barriers to Implementation

Despite the clear need for climate adaptation and sustainability measures, several barriers can hinder their successful implementation.

Financial Constraints

One of the most significant barriers to implementing climate adaptation projects is the lack of sufficient funding. Many cities struggle to secure the necessary financial resources to invest in infrastructure upgrades, green spaces, and other adaptation measures. High upfront costs and competing budget priorities often limit the scope and scale of such projects.

Political and Institutional Challenges

Political will and institutional capacity are critical for the success of climate adaptation initiatives. However, political instability, short-term planning horizons, and bureaucratic inertia can impede progress. Institutional fragmentation, where responsibilities are divided among multiple agencies with limited coordination, further complicates implementation efforts.

Technical and Knowledge Gaps

Implementing effective climate adaptation measures requires access to advanced technical knowledge and expertise. Many cities, especially those in developing countries, face significant technical and knowledge gaps. These include a lack of data, inadequate planning tools, and limited capacity to assess and manage climate risks effectively.

Public Awareness and Engagement

Low levels of public awareness and engagement can also pose a barrier to implementation. Without a clear understanding of climate risks and the benefits of adaptation measures, communities may resist changes or fail to support necessary initiatives. Engaging the public and building broad-based support is crucial for the success of adaptation efforts.

Opportunities for Implementation

Despite these challenges, there are numerous opportunities to advance climate adaptation and sustainability initiatives in urban areas.

Innovative Financing Mechanisms

Exploring innovative financing mechanisms can help overcome financial constraints. Public-private partnerships (PPPs), green bonds, and climate finance from international organizations provide alternative sources of funding. Leveraging these resources can enable cities to invest in large-scale adaptation projects and infrastructure improvements.

Strengthening Institutional Capacity

Building institutional capacity and fostering political will are essential for effective implementation. Training programs, knowledge exchange initiatives, and capacity-building efforts can enhance the ability of local governments to plan and execute adaptation measures. Encouraging interagency collaboration and integrating climate considerations into all aspects of urban planning can also improve outcomes.

Leveraging Technology and Data

Advances in technology and data analytics offer significant opportunities for improving climate adaptation efforts. Geographic information systems (GIS), remote sensing, and predictive modeling tools can enhance risk assessment and decision-making processes. Investing in technology and data infrastructure can empower cities to implement more targeted and effective adaptation strategies.

Community Involvement and Education

Engaging communities and raising public awareness about climate adaptation is crucial for building support and ensuring the success of initiatives. Public education campaigns, participatory planning processes, and community-driven projects can foster a sense of ownership and commitment to sustainability goals. Empowering residents to take an active role in adaptation efforts can lead to more resilient and cohesive communities.

Conclusion

As we have explored throughout this book, the intersection of technology, innovation, and climate adaptation plays a critical role in shaping the resilient and sustainable cities of the future. This conclusion synthesizes the key insights from each chapter, identifies promising directions for future advancements, and calls on policymakers, practitioners, and citizens to take decisive action.

Summary of Key Points

In our journey through the complexities of urban climate adaptation, we have examined numerous strategies, technologies, and frameworks that are essential for building resilient cities. From understanding the fundamental challenges posed by climate change to exploring cutting-edge solutions, the chapters provided a comprehensive overview of how urban areas can adapt and thrive in the face of environmental threats.

The Role of Cities in Climate Change

Urban areas are at the forefront of climate change impacts due to their high population densities and significant carbon footprints. Effective urban planning and the adoption of sustainable practices are crucial for mitigating these impacts.

Smart Cities and Climate Resilience

Smart city technologies, including IoT and Big Data, offer powerful tools for enhancing urban resilience. These technologies enable real-time monitoring, efficient resource management, and proactive responses to climate-related challenges.

Renewable Energy Innovations

Transitioning to renewable energy sources is a cornerstone of urban climate mitigation. Innovations in solar, wind, and hydropower technologies are making clean energy more accessible and affordable, reducing dependence on fossil fuels.

Green Infrastructure and Urban Sustainability

Incorporating green infrastructure, such as green roofs, urban forests, and sustainable drainage systems, enhances urban ecosystems and reduces the urban heat island effect. These natural solutions contribute to improved air quality, biodiversity, and overall urban resilience.

Transportation and Mobility Solutions

Electrifying public transport, promoting active transportation, and implementing smart traffic management systems are key strategies for reducing urban emissions and improving mobility. Integrating these solutions into urban planning can significantly decrease the carbon footprint of cities.

Building Smart and Sustainable Cities

Designing energy-efficient buildings and retrofitting existing structures with sustainable technologies are essential for reducing energy consumption and greenhouse gas emissions. Smart building technologies further enhance efficiency and occupant comfort.

Data-Driven Urban Planning

Utilizing predictive analytics and data-driven approaches enables cities to anticipate and respond to climate impacts more effectively. This proactive planning reduces vulnerabilities and enhances urban resilience.

Climate Adaptation Strategies

Effective climate adaptation strategies encompass a range of measures, from flood management and coastal protection to urban heat island mitigation and community engagement. Implementing these strategies requires a coordinated effort across all levels of government and society.

Financing and Policy Frameworks

Securing adequate financing and establishing supportive policy frameworks are critical for driving climate adaptation efforts. Innovative funding mechanisms, public-private partnerships, and robust regulatory frameworks can accelerate the implementation of sustainable solutions.

Future Directions for Technology and Innovation in Urban Climate Mitigation

As we look to the future, continued advancements in technology and innovation will be paramount in enhancing urban climate mitigation efforts. Several key areas hold particular promise for driving progress and ensuring that cities can adapt to and mitigate the impacts of climate change effectively.

Advancements in Renewable Energy

Future innovations in renewable energy technologies will focus on increasing efficiency, reducing costs, and improving energy storage solutions. Breakthroughs in solar panel efficiency, offshore wind technology, and battery storage will make renewable energy more viable and widespread.

Smart City Technologies

The next generation of smart city technologies will leverage artificial intelligence (AI) and machine learning to optimize urban systems further. AI-driven analytics can predict and manage energy demand, traffic flows, and environmental conditions, enhancing the overall efficiency and resilience of urban areas.

Climate-Resilient Infrastructure

Developing infrastructure that can withstand extreme weather events and changing climate conditions will be crucial. Innovations in materials science, such as self-healing concrete and climate-resistant building materials, will play a significant role in creating durable and adaptable urban infrastructure.

Integrated Urban Planning

Future urban planning will increasingly integrate climate adaptation and mitigation strategies into a cohesive framework. Advanced simulation tools and collaborative platforms will enable planners to design cities that are not only sustainable but also resilient to future climate scenarios.

Community-Centered Approaches

Empowering communities to participate in climate adaptation efforts will become even more critical. Leveraging digital tools and social media platforms can enhance community engagement, education, and collaboration, ensuring that adaptation measures are inclusive and effective.

Call to Action for Policymakers, Practitioners, and Citizens

The journey towards creating resilient and sustainable cities requires collective action and a shared commitment to implementing innovative solutions. Policymakers, practitioners, and citizens all have essential roles to play in this transformative process.

For Policymakers

Develop and enforce robust policy frameworks that support climate adaptation and mitigation efforts. Invest in research and development to drive technological innovation and provide financial incentives for sustainable projects. Foster international collaboration to share knowledge and resources, and ensure that adaptation strategies are inclusive and equitable.

For Practitioners

Embrace the latest technologies and best practices in urban planning and design. Collaborate across disciplines to create integrated solutions that address the multifaceted challenges of climate change. Advocate for sustainable practices within your organizations and communities, and contribute to building a culture of resilience.

For Citizens

Stay informed about the impacts of climate change and the importance of adaptation measures. Participate in local initiatives and advocate for sustainable practices in your community. Adopt sustainable behaviors in your daily life, such as reducing energy consumption, using public transport, and supporting green spaces. Your actions can drive significant change and inspire others to join the effort.

The path to urban resilience and sustainability is challenging, but it is also filled with opportunities for innovation and growth. By working together, we can create cities that not only survive but thrive in the face of climate change. Let us commit to this journey with determination and a shared vision for a better future.

www.ingramcontent.com/pod-product-compliance
Lightning Source LLC
Chambersburg PA
CBHW071921210526
45479CB00002B/508